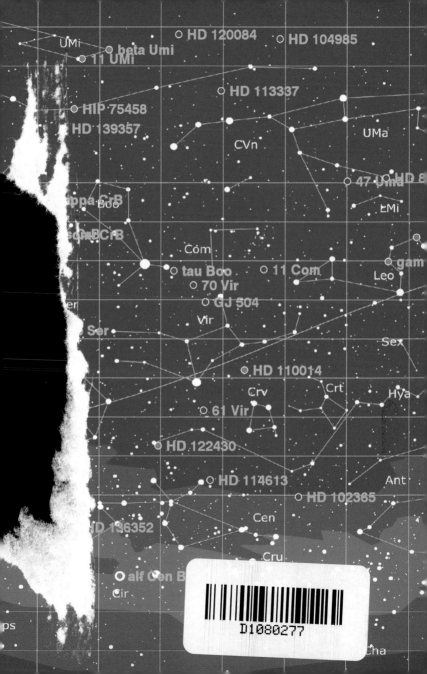

Stargazing

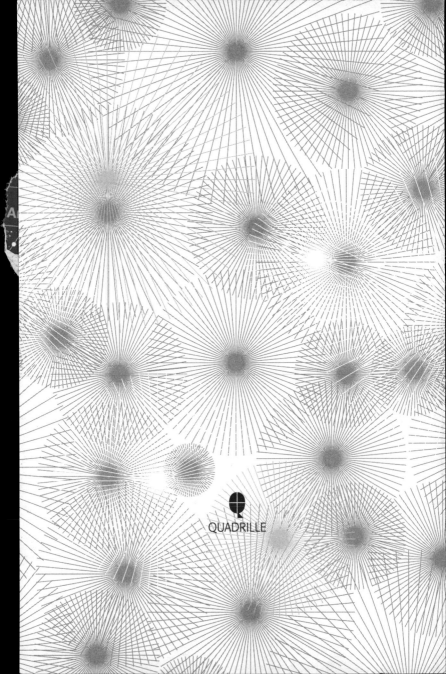

QUADRILLE

The knowledge.
Stargazing | Maggie Aderin-Pocock, MBE

I dedicate this, my first book, to the most important people in my life Lori and Martin. Thank you both for your love, support and encouragement.

Publishing consultant Jane O'Shea
Editor Romilly Morgan
Creative director Helen Lewis
Art direction & design Claire Peters
Illustrator Claire Peters
Production Vincent Smith, Tom Moore

First published in 2015 by
Quadrille Publishing
www.quadrille.co.uk
Quadrille is an imprint of Hardie Grant.
www.hardiegrant.com.au

Text © 2015 Maggie Aderin-Pocock
Design and layout © 2015
Quadrille Publishing

Moon through binoculars page 88 ©
2015 Sky and Telescope
Constellation map page 154 and
endpapers: PHL @ UPR Arecibo, Jim
Cornmell (phl.upr.edu)

The rights of the author have
been asserted.

Cataloguing in Publication Data:
a catalogue record for this book
is available from the British Library.

ISBN 978 184949 621 6

Printed in the UK

1

HISTORY OF STARGAZING

Astronomy is, to my mind, the oldest science by far. I think that this is because it is so accessible, and great inroads in our understanding of the universe have been achieved just by using our eyes.

A concise definition of astronomy is, 'The branch of science which deals with celestial objects, space, and the physical universe as a whole'.* But I like to summarise it as the study of all things not of this Earth, which covers pretty much the whole of the universe.

My favourite quote on the subject comes from Plato, who said, 'It is clear to everyone that astronomy at all events compels the soul to look upwards, and draws it from the things of this world to the other'. To me this sums up the joy of astronomy and the sheer visceral delight that can be gained by the simple act of looking up. That accessibility continues today, for modern astronomy is one of the few areas of science where the amateur can make a real contribution to the subject. With so many astronomers, professionals and amateurs alike, actively watching the skies from many different viewpoints across the world, it can be the amateur enthusiast who notices an event first and then alerts the professionals to observe it with their larger telescopes.

---

* *Oxford English Dictionary* definition

ARCHAEOASTRONOMY

In recent years I have become interested in the subject of archaeoastronomy, which is the study of the astronomy of ancient cultures. What really strikes me about the subject is the fact that nearly every culture has had an interest in astronomy. Spanning time and distance, it would seem that most ancient peoples looked up and formed ideas about what they could see.

Much of the astronomy conducted by the ancients was fuelled by a desire to understand the gods better, to help with navigation to obtain more accurate timekeeping, and by a strong wish to predict the future. Some cultures give us insight into their astronomical past through the records they kept, while for other cultures, such as the Australian Aboriginals, who pass their knowledge and histories by word of mouth, our understanding has come through the work of anthropologists. The Australian Aboriginals have a rich history of stargazing, but rather than naming constellations just from the stars visible in the night sky, they also incorporated dark shadow clouds of dust and gas (nebulae) in our galaxy, the Milky Way, to form a 'constellation' they call 'the Emu in the Sky'.

Some of the earliest astronomical records date back to the 4th century BC when the Chinese recorded and named many stars. By 750 BC the Babylonians had created the first almanacs detailing the movements of the Sun, Moon and planets for astrological purposes.

ANCIENT GREECE

Although some early Greek astronomers, such as Aristarchus of Samos (c. 310–c.230 BC), posited the idea of a heliocentric system in which the Sun is at the centre and the Earth and planets revolve around it, by 140 AD the influential ancient astronomer Claudius Ptolemy (c. 90–c.168 AD) had firmly positioned himself in the Earth-centred camp. His treatise on astronomy, the *Almagest*, describes his geocentric (Earth-centred) model of the cosmos, known as the Ptolemaic system, and includes his own catalogue of the stars, complete with 48 constellations. The geocentric model seemed irrefutable at the time. People could see the evidence for this idea with their own eyes. The Sun rose in the east and set in the west as it journeyed around the Earth, and the stars appeared to travel around the Earth too – it all seemed to make sense. But there were a few dissenters: people who looked at the heavens closely and noticed that there were objects that did not move like the other bodies. They called these 'wandering stars' – the seven visible planets in our solar system – as their motion through the night sky seemed a little erratic.

The *Almagest* was to be a huge influence on astronomers and their thinking for over a thousand years until the early Renaissance, when Copernicus formulated his heliocentric theory of the universe (page 11).

STARGAZING IN INDIA

Interestingly, while Europe was still thinking of the Earth as the centre of the universe, in 499 AD the Indian mathematician and astronomer Aryabhata was considering the idea of the heliocentric solar system, where the planets spin on their axis and follow elliptical orbits around the Sun. He also describes how the Earth rotates about its axis, causing day and night, and also orbits the Sun in a yearly cycle. He calculated that the Earth's diameter was 13,383km (8,316 miles) – today's measurement is 12,742km (7,917 miles) – and this measurement remained the most accurate approximation for over a thousand years.

WESTERN DEVELOPMENTS IN ASTRONOMY

In Europe it took the calculations of Polish astronomer Nicolaus Copernicus (1473–1543) in the 16th century to move away from the Ptolemaic system. His close astronomical observations strongly suggested a heliocentric model of the universe; with this knowledge, the wandering stars now made sense – they were planets like the Earth, orbiting the Sun.

Around this period the telescope was invented, not by Galileo, but by a Dutch optician called Hans Lippershey (1570–1619). Although Lippershey applied for a patent for his invention, it is thought that it was not granted because many similar devices were around at the time.

The flamboyant Danish nobleman Tycho Brahe (1546-1601) lived a rather colourful life but made very detailed measurements of the stars, planets and comets. Although Brahe did not subscribe to the Copernican view of the universe, the measurements that he took were later used by astronomers such as Johannes Kepler (1571-1630) to work out his three laws of planetary motion. The main one of these indicated that planets travelled around the sun in ellipses rather than circles, with the Sun sitting at one focus of the ellipse. These laws of motion are still used by scientists today.

Galileo Galilei (1564–1642), one of the greatest scientific minds of the Renaissance period, was the first person to use a telescope to observe celestial objects. He made a number of the devices, ranging in magnification from x3 to x30 as he improved the design. With them he was able to observe the craters of the moon, see the rings of Saturn and observe the moons of Jupiter. He also discovered that Venus exhibited phases similar to those of the Moon. This turned out to be a giant leap for astronomy as now the heavens could be viewed in more detail than ever before.

The telescopes that Galileo made were refracting devices, where light passes through a lens to gain magnification. Galileo mentioned in his detailed notes the idea of using a reflecting surface to give far better images than trying to view objects through the distorted, bubbled glass available at the time – an early idea for the reflecting telescope. The Italian scientist-priest Niccolò Zucchi (1586–1670) went on to make a reflecting telescope using a spherical surface for the magnification, in around 1616. Although the images would have been distorted, it seems he used the device to record the bands on Jupiter's surface.

English physicist Isaac Newton (1642–1726) was the first scientist to make a reflecting telescope that actually worked, now known as the Newtonian telescope. He used his telescope to observe the universe and form his ideas of the universal law of gravity, which is the force of attraction that holds the universe together.

DEVELOPMENT OF THE MODERN TELESCOPE

From the time of Newton and his Newtonian telescope, lens and mirror telescopes got bigger and better over the centuries, but pockets of turbulent air in the atmosphere were a major limitation. The size of these atmospheric pockets is of the order of 4m (13ft) – if you make your telescope much larger than this, the levels of distortion increase greatly. The only answer was to have an observatory in space, something first proposed by American astronomer Lyman Spitzer in 1946. A number of astronomical observatories were launched starting in the 1970s, culminating in the launch of the Hubble Space Telescope in 1990. Although space telescopes are very effective, they are also very expensive.

ADAPTIVE OPTICS

The problem of turbulence for terrestrial-based observatories was resolved by a system developed by the military in the 1990s called adaptive optics. American astronomer Horace Babcock (1912–2003) first proposed this method in the 1950s. Adaptive optics uses a system that measures the distortion caused by atmospheric turbulence, then corrects the image and removes the distortions. This is done by changing the shape of a mirror or the optical density of a lens in the telescope to counter the shape of the distortion.

Thanks to adaptive optics, telescopes have grown to the 10m (33ft) giants that we see today, and adaptive optics systems sit on all the larger telescopes. However, this has not negated the need for space telescopes (see pages 138–9).

With the use of the telescope as an astronomical tool, our knowledge of the universe has been transformed. From the earliest theories of an Earth-centred universe to those that embraced the heliocentric model, our ideas of the universe have evolved and continue to evolve. The real beauty of astronomy is that, just like our ancestors, it is an activity we can all participate in as we contemplate what lies beyond.

DISTANCES IN THE UNIVERSE

We all know that the universe is very big, so describing distances using Earthly units of kilometres and miles soon gets very cumbersome. So over the years astronomers have come up with a number of units that cut down the number of zeros we have to add when talking about the larger distances.

Astronomical unit: Average distance between the Earth and the Sun taken to be 149,595,871km or around 93,000,000 miles.

Light year: This is actually a distance rather than a unit of time and is the distance light travels through in a vacuum in a year: 9.4607×10^{12} that is around 9.5km (with 12 zeros after it) or around 6 million, million miles.

A parsec: This is a much larger distance equivalent of 3.26 light years. Its definition is a bit more complicated. One parsec is the distance at which one astronomical unit subtends an angle of one arcsecond (one arcsecond is 1/3600th of a degree). The parsec is a useful measure for professional astronomers as a way of measuring the parallax on a star from two widely spaced distances. We can then use geometry to work out the distance to the star.

To get a feeling of how truly immense the universe is, look at the distances in the table below of a few well-known celestial objects. Let us start locally in our solar system.

Distance	Time taken for light to travel distance
Earth to the Moon	1.255 seconds
Earth to the Sun	8.3 minutes
Sun to Jupiter	41 minutes
Sun to Saturn	85 minutes
Sun to Neptune	4.2 hours
Sun to Voyager I	17.1 hours

To state the distance in another way, it currently takes 17 hours for us to receive a signal from Voyager. Let's move further out...

Distance	Time taken for light to travel distance
Sun to Alpha Centauri (our closest star)	4.3 years
Sun to Sirius (our brightest star)	8.6 years
Sun to 61 Cygni	11.4 years
Sun to Pollux	33.7 years
Sun to Castor	51.6 years
Sun to Regulus	77.6 years

From the distance of Regulus our Sun would barely be visible with the naked eye.

Distance	Time taken for light to travel distance
Sun to Acrux (the brightest star in the Southern Cross)	321 years
Sun to the Pleiades	385 years
Sun to Polaris	432 years
Sun to Betelgeuse (In constellation Orion)	640 years
Sun to Rigel (In constellation Orion)	777 years
Sun to the Orion Nebula (In constellation Orion)	1,300 years
Sun to the Crab Nebula	6,300 years
Sun to the Double Cluster	7,200 years
Sun to Omega Centauri	16,300 years
Sun to M13	21,000 years
Sun to the Galactic centre	27,700 years
Galactic diameter (time for light to travel across own galaxy)	81,500 years

At this distance we have reached the limit of our galaxy!

Distance	Time taken for light to travel distance
Sun to the Large Magellanic Cloud (the nearest small galaxy)	160,000 years
Sun to the Andromeda Galaxy (the nearest large galaxy)	2,540,000 years

Beyond this distance we need to use equipment to observe the universe, as we have now left the region visible with the naked eye.

Distance	Time taken for light to travel distance
Sun to UGC 8091 (the farthest galaxy of our galaxy cluster)	7,900,000 years
Sun to M81 (spiral galaxy in Ursa Major)	12,000,000 years
Sun to M104 (the Sombrero Galaxy)	30,000,000 years
Sun to M87 (a spherical galaxy in Virgo)	55,000,000 years

At this stage, light reaching us from the next series of bodies actually started its journey before the extinction of the dinosaurs some 65 million years ago.

Distance	Time taken for light to travel distance
To Perseus Group (a group of 500 galaxies in Perseus)	190,000,000 years
To Coma Group (a group of 1000 galaxies in Coma Berenices)	225,000,000 years
To Hercules Group (a group of galaxies in Hercules)	350,000,000 years
To Boötes Group (a group of 150 galaxies in Boötes)	1,240,000,000 years
To 3C273 (the first quasar discovered)	2,000,000,000 years
To Q 0134+329 (typical quasar)	4,500,000,000 years
To Galaxy 970228 (typical Gamma Ray Burster found in 1997)	6,000,000,000 years
To GRB 090429B (furthest Gamma Ray Burster discovered in 2011	13,000,000,000 years
To remotest quasars discovered in 2011	29,000,000,000 years
To remotest protogalaxy discovered in 2011	32,000,000,000 years
Edge of universe (limit of observable universe)	47,000,000,000 years

2
CELESTIAL
MECHANICS

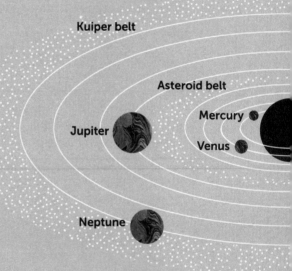

Oort Cloud

Kuiper belt

Asteroid belt

Mercury

Jupiter

Venus

Neptune

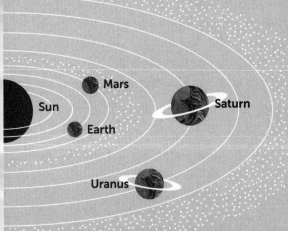

Our understanding of our place in the universe has been evolving over many centuries, both with the use of naked-eye observations and more recently with the aid of increasingly sophisticated scientific equipment.

To gain an awareness of what we can actually view in the universe, it's helpful first to get a sense of what is out there. Let's take a tour, travelling out from our inner solar system to the edge of known space. Not all the objects mentioned in this section are viewable from Earth, even with the best telescopes, but knowing what lies in the heavens will gives us a better understanding of our universe and our place in it.

THE PLANETS

We're all familiar with our home planet, Earth, but this is just one of eight planets that are in orbit about the Sun.

MERCURY

The closest planet to the Sun is Mercury, named after the Roman messenger of the gods, which is the smallest planet in our solar system and is a lot hotter than Earth (its daytime temperature can reach 427°C/800°F). It doesn't have a moon. Not much is known about Mercury – it's thought that if it did ever have an atmosphere, it would have been torn away by the solar wind (charged particles that flow from the Sun's surface).

We will soon get our first close-up glimpse of Mercury with the European Space Agency space mission, Bepi Columbo, due to be launched in 2017.

VENUS

The second planet from the Sun is Venus, and is seen as one of Earth's twins in terms of size and mass – but there the similarities end, as Venus is a fiery planet covered in possibly active volcanoes with scorching temperatures. Most of us will have observed Venus, but many may not have recognised it as a planet. Due to its closer orbit of the Sun, Venus is often seen as a very bright 'star' just before sunrise or just after sunset. Historically, Venus was associated with beauty and love – but it was not until 1970, when the Russian Venera 7 robotic space probe landed on its surface, that we discovered a barren planet with a corrosive atmosphere. After a hard impact on landing due to a malfunctioning parachute, the probe valiantly transmitted for 23 minutes before succumbing to the inhospitable atmosphere, with a temperature of around 467°C (872°F), and pressure 90 times that of the Earth's surface. Venus is even hotter than Mercury, despite the fact that it lies further from the Sun, because the Venusian atmosphere is 96 per cent carbon dioxide*. Like Mercury, Venus does not have any moons.

---

* A greenhouse gas found on Earth that traps infrared radiation from the Sun in the atmosphere, elevating the temperature.

EARTH

The Earth is the third planet from the Sun, and has a single moon in orbit about it. From space it looks like a beautiful blue marble (in fact, this is the title of a famous photograph taken of Earth from the Apollo 17 spacecraft in 1972), thanks to its vast oceans (water covers about 71 per cent of the Earth's surface). It is thought that it was an image like this that triggered the environmental movement, suddenly seeing our planet alone and vulnerable in space. The Earth is the densest planet in our solar system, but only the fifth largest in terms of diameter, at 12,742km (7,918 miles).

MARS

Beyond Earth lies the planet Mars, named after the Roman god of war, with its two moons, Phobos and Deimos (the Greek names for the twin sons of Venus and Mars; not surprisingly, their names are the ancient Greek for fear and terror). Mars is the fourth planet from the Sun, and the second smallest planet in our solar system. It is seen by many scientists to be the Earth's twin in terms of its environment, as its temperature of -56°C (-70°F) is the closest of any planet in our solar system to that of the Earth's. Scientists also think Mars may have had a large ocean at one time, which is another similarity with the Earth. Mars lies in the cusp of a so-called Goldilocks Zone, which is a potentially habitable region of space that broadly replicates the relationship between our own Sun and Earth; therefore the conditions of the planet (or planets) can support water on its surface. Evidence studied by planetary geologists indicates that Mars did indeed once have water flowing over its surface, but this is no longer the case. Water found on the planet is frozen into the ground or sits as surface ice at the northern and southern polar caps. Why the atmosphere of Mars changed so radically remains a mystery, but it is one of the questions that the many exploration orbiters and rovers on and around the planet have been sent to find an answer to.

JUPITER

At this distance from the Sun we enter the region of the gas giants, Jupiter and Saturn, so-called because they are composed of helium and hydrogen. Jupiter, named after the Roman king of the gods, is the first that we encounter. It is the fifth planet from the Sun and the largest planet of our solar system (at a whopping 142,797km (88,730 miles) in diameter it is just over 11 times the size of the Earth, and in terms of volume you can fit 1,000 Earths into one Jupiter). Jupiter can be seen as Earth's defender: as thanks to its size and mass, it helps protect the Earth by gravitationally attracting objects that could otherwise hit our planet. Jupiter has around 65 moons in orbit about it.

SATURN

Beyond Jupiter lies Saturn, named after the Roman god of agriculture, a beautiful planet surrounded by vibrant rings, which are made of tiny particles of mainly water ice. It is the sixth planet from the Sun, and the second-largest planet in our solar system. Scientists have been learning more about this mysterious, gaseous planet through the Cassini-Huygens space mission. Saturn has over 50 confirmed moons, but more potential candidates are being investigated. The Huygens space probe landed on one of the larger moons, Titan, and sent back amazing images of the surface.

URANUS AND NEPTUNE

The two outer planets of our solar system are Uranus and Neptune, sitting seventh and eighth respectively. Uranus was the Roman god of the sky, while Neptune was the god who ruled the seas. Often known as the ice giants because of their icy composition, these sentinels of our solar system sit a long way away from where we live and not much is known about them. They have mainly been studied via the Voyager spacecraft and recently the Hubble Space Telescope. Uranus is the third largest planet in our solar system, and Neptune is the fourth largest.

BELTS AND CLOUDS

As well as planets, there are many other objects orbiting the Sun, most notably the Asteroid Belt, the Kuiper Belt, and the curious sounding Oort Cloud (named after the man who discovered it in 1950, the Dutch astronomer Jan Oort).

ASTEROID BELT

Just beyond Mars lies the asteroid belt. Material in this region comes in many different shapes and sizes. It consists of billions of asteroids, which range from pieces of very large rock (some of the largest are around 240km/150 miles in diameter) to clumps of rubble held together by gravity. They are thought to be part of the detritus left behind during the formation of the main planets. The gravitational pull of Jupiter is thought to have stopped these objects from forming a planet. However, the asteroid belt has one dwarf planet, Ceres. A dwarf planet is an astronomical object that orbits a sun, is large enough to be approximately spherical in shape, owing to its gravitational mass but, unlike a planet, it has not cleared all the material in its orbit. To date scientists have discovered five confirmed dwarf planets in our solar system – Pluto, Ceres, Haumea, Makemake and Eris – but there may be as many as 50 more out there. In early 2015, Ceres caused excitement when the Dawn Space Probe took pictures of its surface, revealing two intriguing bright spots. Scientists think they may be sunlight reflected from patches of surface ice, but no one knows yet.

KUIPER BELT

Similar to the Asteroid Belt but much wider and more massive, the Kuiper Belt sits out beyond Neptune. It is named after the

---

* Pluto was reclassified due to its small size.

Dutch-American astronomer Gerard Kuiper (1905–1973). The largest object in the Kuiper belt is Pluto, which was thought to be a planet but has been recently reclassified. The Kuiper Belt is also thought to be the origin of the comets we see in our night skies. Comets are icy bodies made of water, rock and carbon-based materials known as organics. Every so often a comet gets nudged out of position in its orbit in the Kuiper Belt by the interactions of the giant planets. It is then sent into an elliptical orbit and is eventually pulled in towards the inner solar system by the Sun's gravity. As the comet draws closer to the Sun, it starts to melt, releasing dust and gases that form a comet's impressive tail.

OORT CLOUD

Out beyond the Kuiper belt lies the Oort Cloud. This huge spherical cloud consists of up to 2 trillion icy bodies that stretch out to a distance nearly a quarter of the way to our Sun's next-door neighbour Proxima Centauri, some 4.24 light years away. This cloud marks the limit of the gravitational reach of the Sun and therefore it can be considered to be the edge of our solar system.

INTERSTELLAR SPACE

We now enter a zone of space that lies between the stars. This region is called interstellar space. Out here there is not very much, but we do find occasional volumes of dust and gas. If these volumes are large and active enough, they are called molecular clouds, or nebulae. Stars are born in these clouds. If undisturbed, a nebula can remain stable, but if it is disturbed by a nearby gravitational event, then its matter starts to clump together due to the gravitational attraction between the particles. If the density of the material becomes high enough, a star is formed, sometimes with a planetary system around it.

THE STARS

Moving further out into space, the next thing we encounter are the stars we see in the night sky, similar to our own Sun. Some are larger and some smaller, some older and some younger. Stars, like people, go through life cycles and, just like people, their appearance changes at the different stages of that cycle. The life cycle of a star depends on its mass: i.e. the amount of stuff that it is made of. The more massive the star is the shorter its life cycle will be. Our Sun, which has an age of about 5 billion years, is of average size and is likely to be around for at least another 10 billion years. A star 10 times the mass of our sun is likely to exist only 30 million years and a star 50 times the mass of the Sun will only survive for around 5 million years.

Stars shine brightly due to a process called nuclear fusion that is happening at their core. This is when atoms fuse together to form new elements. Most stars start off as a gaseous nebula made up of mainly hydrogen and dust particles, which are drawn together by gravitational forces. Once the core of the infant star is dense enough, hydrogen atoms fuse together to make helium through nuclear fusion. This fusing process releases amazing amounts of energy and stops gravitational forces from making the star collapse in on itself.

The energy released from the surface can be detected as various forms of radiation. One type of radiation is visible light, which we can see with our eyes, and this is the reason why we can see stars as bright objects. Other energy released includes infrared, ultraviolet, radio waves and X-rays, to name but a few.

THE SUN

The life cycle of a star is very dependent on the fusion that is taking place at its core. Take our Sun – it's about halfway though its life cycle and is consuming the hydrogen, leaving behind a growing core of denser helium. As the Sun continues to age the hydrogen will become more and more depleted and the helium core will start to collapse in on itself. When this happens the pressure at the core will increase. At some point the helium core

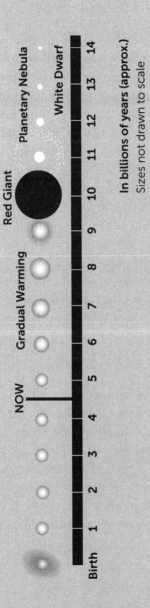

THE LIFE CYCLE OF THE SUN

Birth 1 2 3 4 5 6 7 8 9 10 11 12 13 14

NOW

Gradual Warming

Red Giant

Planetary Nebula

White Dwarf

In billions of years (approx.)
Sizes not drawn to scale

33

will reach a critical pressure where further fusion of helium into other heavier elements is possible. This causes a release of energy that forces the outer layers of the Sun to expand outwards and cool. At this stage the Sun will become a red giant, so-called because the appearance of a red giant is a luminous red. The expansion of these layers will eventually become large enough to subsume all the planets of the inner solar system (Mercury, Venus, Earth and Mars), but don't worry: this will not happen for many billions of years!

While the outer regions of the Sun continue to expand, the helium nuclei in the core will continue to fuse into carbon. Now, when carbon is made, the pressure in the core is not enough to convert it into heavier elements, so no further fusion is possible. The Sun's core will stabilise, but the outer layers will continue to expand and will eventually sit independently of the core. At this stage the core will form a white dwarf planet, and the now free outer layers will form a planetary nebula. This marks the final stage of the Sun's life cycle.

A white dwarf is the final phase in the life cycle of a star the size of our Sun, but for larger stars the end can be a lot more dramatic. Stars some 10 to 100 times bigger than our Sun can continue fusion past the carbon phase and make heavier elements such as neon, oxygen, silicon and eventually iron. For these massive stars, their final stages result in a huge explosion of energy known as a supernova. This release is so bright that we can detect supernova events occurring in distant galaxies.

Left behind in the core of a supernova can be either a neutron star or a black hole.

NEUTRON STARS

A neutron star is an object with a mass a few times more than that of our Sun, but with a radius of only 10km (6 miles). This object is so super-dense that a teaspoon of this matter on Earth would weigh as much as a mountain!

BLACK HOLES

The alternative to a neutron star, a black hole, is one of the great enigmas of our universe. This is a body so dense, and with such a super-powerful gravitational field, that not even light can escape from its gravity. As light is not emitted from this region it should make it very hard to detect, but this is not the case. Because of their strong gravitational field black holes disturb the path of many neighbouring bodies, providing us with clues to their location.

EXOPLANETS

One of the most exciting developments in astronomy in recent years has been the detection of extrasolar planets or exoplanets. These are planets that are in orbit about distant stars. To date, approximately 2,000 exoplanets have been detected and scientists are expecting to find more. One of the questions that arises in the study of exoplanets is, 'Is there life out there?' So far, no life has been found, but with more of these planets being discovered, and with better telescopes available in the future, it may be possible to find indications of life on another planet.

GALAXIES

When we encounter stars in the universe we are most likely to find them in large clusters called galaxies. Our solar system exists in a galaxy called the Milky Way, which contains approximately 200 billion stars. All of the stars that we're able to see in any detail belong to the Milky Way. Galaxies vary in size and shape; one of the most common shapes is the spiral. The Milky Way is a spiral galaxy and our Sun sits in one of its spiral arms.

The cluster of stars that make up our galaxy can be seen quite clearly in the northern night sky if the night is very clear and there is little light pollution around. It's much easier to spot the Milky Way, in the skies of the southern hemisphere, as from this viewpoint you look straight into the centre of the galaxy where there is a greater abundance of stars. The Milky Way, as the name suggests, looks like a broad, silvery or 'milky' pathway across the darker night sky.

THE UNIVERSE

Finally, to end our journey through the firmament we need to discuss the universe. The universe contains all space, time, matter and energy. Scientists currently believe that the universe began some 13.8 billion years ago and started with something that we call the Big Bang. The theory of the Big Bang was postulated to explain the observed phenomena that most of the visible stars and galaxies in the universe seem to be moving away from us. Logically, if they are moving away from us now, going back in time they must have all converged in a single point from which they originated – the Big Bang. Although the Big Bang idea is still just a theory, it has stood the test of time by best fitting the observations scientists have made to date.

So this is our playground, the universe. In the next chapters we will get an understanding of what we can observe and the best ways to do it.

MYSTERIES AND WONDERS OF THE UNIVERSE

Big Bang: Edwin Hubble made observations of distant galaxies and noticed that the further they were from us, the faster they seemed to be travelling away from us. This means that if we reversed time, all things in the universe would coalesce into a single point. From this the theory of the Big Bang was formed, suggesting that the universe started from a single point and then expanded.

Dark energy: When we observe the universe and its expansion we have noticed that it is expanding faster than expected. The cause is unknown but it has been given the term dark energy to describe the force that is expanding the universe. We currently believe that dark energy accounts for 68 per cent of the universe.

Dark matter: When we observe the movement of large bodies in space, such as galaxies, we notice that their movement isn't consistent with the matter that we can see within that body. Many of the galaxies we observe should fly apart if following the laws of physics. As we believe that the laws of physics are universal, we need some other explanation of what is holding these bodies together. Nothing has yet been found but we call the invisible matter that we think is there but can not see, dark matter, as it does not interact with the electromagnetic spectrum. It is of great importance as it makes up around 27 per cent of the universe.

Black holes: A black hole occurs when a truly massive object collapses in on itself due to its own gravity. The resultant body has such a strong gravitational field that not even light can escape its clutches. Black holes have been observed in space by the effect that they have on other bodies in their surroundings. Although we cannot see them, we can see other bodies rotating around them and sometimes we can see matter from other bodies being pulled into them. It is thought that at the centre of every galaxy (including our own) there is a super massive black hole.

Multiverse: The multiverse is a mind boggling idea that posits that as well as this universe there is a multiplicity of other universes out there. It seems a bit crazy but many of the scientific theories that we are working on today predict the possibility of the multiverse.

Spacetime: Spacetime was a concept first proposed by Einstein's teacher Hermann Minkowski. The idea was to take the three dimensions of space and link these with 'time' as a 4th dimension to form a continuum. Einstein used spacetime in his theory of general relativity to analyse gravity, a curvature in spacetime.

Wormholes: Wormholes are the friend of science-fiction writers. They are thought to significantly cut journey times across space by bending spacetime and were a prediction of Einstein's theory of general relativity, but none have been found yet.

Gravitational waves: Predicted by Einstein's theory of general relativity, a gravitational wave is a ripple in the curvature of spacetime which travels outwards from its source as a wave. These waves carry gravitational radiation. To date no gravitational waves have been detected but a number of space- and ground-based experiments have been set up to detect them.

Cosmic inflation: Through the Big Bang theory we believe that the universe has been expanding. The idea behind cosmic inflation is that the universe went through a period (less than a trillionth of a second) of massive expansion, growing from the size of a sub-atomic particle to the size of a grapefruit. It was used to explain the unevenness of cosmic background radiation and why we have the formation of clumps of matter in the universe.

3

GETTING
STARTED

It's amazing how much of the night sky we can see if the conditions are right and we choose a good location from which to view it. I first became interested in stargazing as a child living in the heart of London, and although the yellow glow of the sodium lamps can swamp many of the stars, much is still visible. In this chapter we explore the basics of stargazing. These tips are applicable to naked eye stargazing or stargazing equipment and provide information on finding a location to view the cosmos as well as describing the coordinate systems that allow us to locate objects in the night sky.

STARGAZING BASICS

For stargazing, clear skies are definitely the best. Visible light is unable to penetrate clouds, so even light cloud will obscure most of the night sky. The average number of clear nights in the UK over the last few years has varied from around 18 per cent to 25 per cent. This sounds quite good, but many of these are likely to occur in the summer, when the longer days mean shorter nights, which will limit the time available to observe.

Of course, stargazing can be performed in any part of the world, so holidays or business travel can provide a great opportunity to go outside and see different views of the universe, sometimes in better weather. The ideal time for stargazing is during the winter months, when the nights are a lot longer than those of summer. For younger stargazers, the sun setting much earlier at this time of year is useful too. However, winter stargazing takes a little planning if it's to be anything other than a quick foray outside to see the skies and then back into the warmth again. As astronomy is by its nature a static activity, layered warm clothes, a peakless hat, gloves and a scarf can make the whole experience much more enjoyable. My feet often seem to be my Achilles heel in terms of the cold (if you'll excuse the pun), so over the years I've invested in some thick walking socks that go a long way to keeping my extremities snug. A flask containing a warm beverage – even just to pour out into a cup and hold – can be very comforting too.

IDEAL VIEWING CONDITIONS

If you have a clear night, you need to optimise your viewing conditions. The main factor that will affect these is light pollution. If you live in a city or large town, positioning yourself away from street lighting is essential. This is often harder than it sounds, but if you can find a clearing that is less well lit, such as playing fields or a park, this is often a good start. When I was at university in London I would often go out into the outskirts of Hyde Park to admire the firmament. However, do be aware of your personal safety when going into less well-lit areas and err on the side of caution; ideally go stargazing with a friend or two, and always let someone know where you're going if it's slightly off the beaten track.

If an open area is not accessible, minimising the number of streetlights visible in your line of sight will definitely help. At home in Guildford, my back garden is relatively free of manmade light pollution, which is mainly emitted from the streetlights at the front of the house. As a child in London I would often go to a small area between our council flats where the street lamps were fewer.

Although the Moon is a wonderful object to observe, it may be a surprise to learn that it can be a source of light pollution. If you're hoping to see some of the dimmer objects in the night sky, it's best to do this near the time of a new Moon or before the Moon has risen or after it has set. Details of the Sun and Moon's rise and set times can be found in a book called an almanac, which can usually be viewed in your local library or much more readily via the Internet, where a search for sunrise/set and moon rise/set will provide you with the information for your location.

DARK ADAPTION

Before you start to observe, you need your eyes to be in a state called 'dark adaption'. In bright light, the pupils of our eyes contract to let less light in, stopping us from being dazzled. If we then move to a dark area our pupils dilate, but this reaction takes some time. So to make the most of a night's viewing, we need to allow our eyes to adjust to the dark by letting our pupils fully dilate. As well as the pupils' response, much of our dark adaption is dependent on the refresh rate of the chemicals in the light receptors located in the retina.

The retina contains two types of light-sensitive cells called rods and cones. Rods work in low light levels, whereas cones require much more light to function. Both rods and cones contain chemicals that break down, or bleach, when exposed to light. Our sensitivity to light is dependent on the amount of the chemical that is present unbleached.

Our eyes have two main modes of operation. In bright light the cone receptors are mainly operational, while conversely the light-sensitive chemical rhodopsin in our rods is bleached. When we encounter low light levels, the rhodopsin starts to regenerate but this can take some time. During this regeneration period, our eyes are not as sensitive to the darker conditions as they could be and as a result we can see fewer stars. The adjustment to low light levels can take around 30 minutes, and during this time it is essential not to be exposed to bright light or the process of dark adaption will be lost. While a torch is a useful tool for stargazing expeditions, care should be taken not to shine it in your own eyes or in anyone else's direction. Eyes that are dark adjusted can respond quite painfully to the full glare of a torch, and even if the flash is short it does take some time for the eyes to recover and see dimmer objects again. Some torches are available with a red filter, or even better a red LED. These are useful because they are less dazzling but can still give some help illuminating one's way.

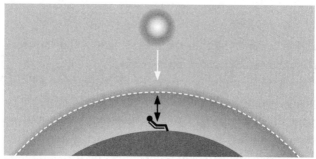

Sun at Zenith*, atmosphere is thinner and better for stargazing

Sun close to Horizon, light takes a longer path through the atmosphere

COMFORTABLE VIEWING

Often people think that stargazing is done standing up; this can be the case, but a chair can be useful too. I find an adjustable deckchair ideal for when I am doing naked-eye observations. I set it up at a shallow-rake angle so that I can sit in it with my head supported looking straight up at the night sky. Also if I

* The zenith angle is the angle between the object under observation and the vertical. When viewing the night sky it is best to observe objects when they are high in the sky, so have a small zenith angle. This reduces the amount of atmosphere that the object is observed through enabling a much less turbulent observation.

wish to change my view, the chair is light and relatively easy to move. I have a friend who uses his children's trampoline to observe. If the sides are not too high, you can get a good view of the night sky while being gently supported by the springs. I've been told that this is comfortable enough to lie on for hours.

NAVIGATING THE STARS

In the following chapters we will be looking at what can be seen first with the naked eye, then a set of binoculars and finally a telescope. Before we plunge into these types of observation it is useful to look at the ways we can locate objects in the night sky.

The brightness of a star as we see it from Earth depends on two things: the intrinsic brightness of the star, and the distance it is away from us. The closer an object is to the observer, the brighter the object will seem. Each of the stars that we see in the night sky is designated with a number called its magnitude, which relates to its brightness in visible light as seen from Earth. Counter-intuitively, the brighter the star is the lower its magnitude will be.

The brightest star we can see in our skies is the Sun and this is allocated a magnitude of -26. A full Moon has a magnitude of -13, while the planet Venus has a magnitude of -4. With the naked eye in a good location away from light pollution we can see an object of magnitude 4. And with a good set of binoculars in good conditions we can see objects of magnitude 8, which is the magnitude of the planet Neptune.

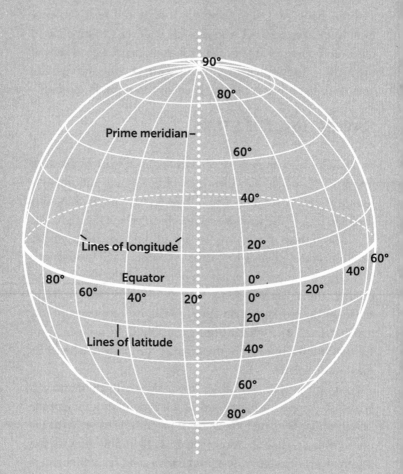

90°

80°

Prime meridian –

60°

40°

Lines of longitude

20°

Equator
80°
60°
40°
20°
0°

0°
20°
40°
60°

Lines of latitude

20°

40°

60°

80°

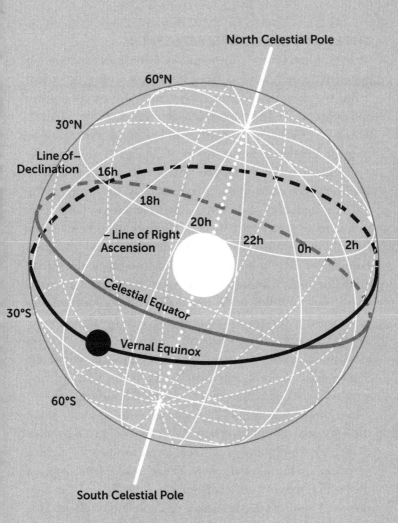

North Celestial Pole

60°N

30°N

Line of—
Declination

16h

18h

20h

—Line of Right
Ascension

22h

0h

2h

Celestial Equator

30°S

Vernal Equinox

60°S

South Celestial Pole

THE EQUATORIAL COORDINATE SYSTEM

To identify a position in the night sky, some form of navigation system is needed. Astronomers generally use an angular system called the equatorial coordinate system, which consists of two components called right ascension (RA) and declination (DEC). This is very similar to the latitude and longitude that we use to pinpoint locations over the surface of the Earth.

The Earth-bound navigation system relies on an imaginary line called the equator, which runs around the centre of the Earth, lying halfway between the north and south poles. This is designated with a latitude of 0°. Both poles lie at 90° away from the equator, so each has a latitude of 90°. So for any given latitude, we have a circle of coordinates around the Earth. To define a position on the Earth's surface we need a second co-ordinate, and this is the longitude. Lines of longitude, or meridians, run from pole to pole, with a prime meridian running through London's Greenwich area. This prime meridian is set at 0°, and locations east or west of this are either plus or minus degrees from the prime meridian point. With this coordinate system we can define any point on the Earth's surface.

In a similar way, we can find objects in the night sky using the equatorial coordinate system. With this system it is possible to identify the position of any star on an imaginary spherical surface called the celestial sphere. The celestial sphere imagines the Earth at the centre of a large sphere, with all the visible stars projected onto the inside surface of the celestial sphere. This way we do not have to worry about the distance of the stars to the Earth and we can allocate the equivalent of a latitude and longitude to each object on the celestial sphere's surface. Within the celestial sphere, latitude becomes right ascension and longitude becomes declination.

However, if you wish to observe a large-scale event, such as a meteor shower, less accurate calculations are needed to define the area for observation. For this type of navigation constellations are often used.

A constellation is a group of stars that can be seen in proximity to each other in a patch of sky. In the past people looked at these stars and imagined they formed images, similar to a dot-to-dot puzzle. Modern astronomers have identified 88 of these star configurations, covering all areas of the sky. These constellations are used as markers to guide people over the sky, as each one denotes a specific location on the imaginary celestial sphere. Well-known constellations include the signs of the zodiac; Orion, the Hunter; Pegasus, the Winged Horse; and the Ursa Major, the Great Bear. It's important to note that although the stars of a constellation may appear clustered together in two-dimensional space on the imaginary celestial sphere, in reality the stars' proximity to Earth may vary even though they appear in the same constellation.

One of the most useful navigation tools that I have been using is a stargazing app. With the built-in GPS of the phone and other information, it can work out what should be around me and help identify constellations, individual stars, planets and other heavenly bodies. I can also use it to locate a point of interest by following arrows that will re-orientate me until I am in front of the object of my desire.

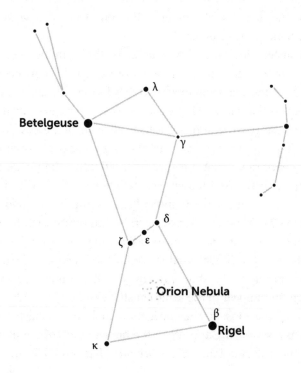

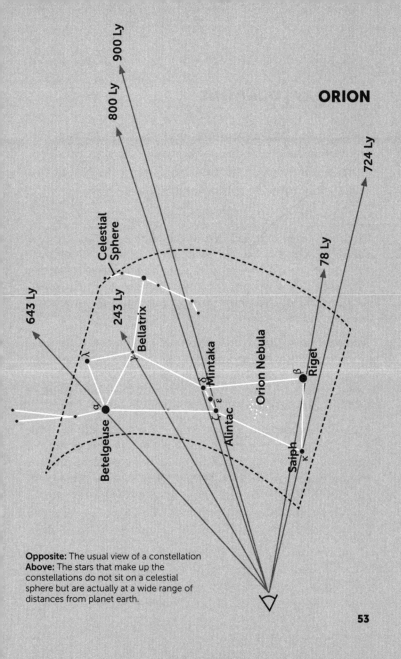

ORION

900 Ly
800 Ly
724 Ly
78 Ly
643 Ly
243 Ly

Celestial Sphere

Bellatrix

λ
γ

Mintaka
δ
ε
ζ
Alintac

Orion Nebula

Betelgeuse
α

β
Riget

Saiph
κ

Opposite: The usual view of a constellation
Above: The stars that make up the constellations do not sit on a celestial sphere but are actually at a wide range of distances from planet earth.

DARK SKY LOCATIONS

Many of the best stargazing locations have already been commandeered by some of the world's largest telescopes. Afterall, why build a huge telescope if you are not going to put it at one of the best stargazing locations in the world?

Uluru/Australian outback: One of my favourite non-professional stargazing spots has been in the outback, especially the secluded location of Uluru. You are so far away from any streetlights that they seem like a distant memory. Also, as with all the southern hemisphere locations, you look into the heart of our galaxy, the Milky Way, so the stars are truly abundant.

Atacama Desert, Chile: This location has the accolade of being one of the driest places on Earth, with this title goes less than one inch of rainfall per year. Many of the future goliaths of the sky will be located here. On my visit, it felt as if I just had stepped onto Mars because of its dry red soil and complete lack of vegetation.

Mauna Kea, Hawaii: As a small island in the middle of the Pacific ocean Mauna Kea gets lots of rain, but if you happen to do your stargazing above the clouds then this is not a problem. This is a glorious location where a number of professional telescopes already sit. At nearly 5000m (16,404ft) the thin air makes you feel a little drunk and quite euphoric but that might have just been the amazing view of the stars.

Namib Rad International Dark Sky Reserve, Namibia: This location is definitely on my bucket list as a must see. It helps that Namibia is one of the most sparsely populated locations on Earth. Couple that with a 1,994 square kiliometres (770 square mile) reserve, and you can see why this location has been described as close to natural darkness as one can obtain in this world. I would dearly love to check this one out for myself.

Canary Islands, Tenerife, Caldera de Taburiente: A location that has been acquired by the professionals and with good reason: just like on Mauna Kea the clouds form below the extinct volcano peak, allowing for amazingly clear skies for most of the year. I have seen my shadow formed by the setting sun projected on the clouds below me and then gone on to have a magical night of observing.

Galloway Forest, Scotland: The forest element of this location doesn't make it sound ideal for stargazing but it is a designated dark sky park consisting of 300 square miles where you can see up to 50 times more stars than you would see in a major city. Also being just a two hour drive from Glasgow proves that dark sky locations don't have to be inaccessible.

4

NAKED EYE
STARGAZING

Although the recorded history of stargazing goes back to around 7000 BC, it is actually even older than that. It is thought that most early humans looked up at the points of light in the sky with wonder, so we are following in the footsteps of our ancient ancestors when we observe the night sky unaided. In fact, it was not until around 400 years ago when Galileo used a telescope to observe the heavens for the first time that this tradition was broken. Unaided observations are also often referred to as naked eye observations.

This chapter explores what we can see in the firmament using just our eyes. I will start us with the brightest objects that can be easily seen and then work our way through to the dimmer, more challenging bodies.

THE SUN AND ECLIPSES

The brightness of an object seen from Earth in visible light is defined as its magnitude (see page 47). Our local star, the Sun, is the brightest object in our skies during the day with a magnitude of -26 and it can be observed, but great caution is needed as it is so bright that it can damage the eyes. Never look directly at the Sun, whether unaided, with binoculars or a telescope.

Naked eye observation of the Sun with the appropriate filtration (such as solar spectacles which reduce the intensity of the light reaching your eyes or with a telescope 'H alpha' filter, which lets a very small range of wavelengths enter the instrument), is rarely done as there is not much detail to see at the level of magnification that is achieved with the human eye. There is one main exception to this, which is the total eclipse of the Sun. To view this you will need special solar eclipse glasses.

A total eclipse of the Sun occurs when the Moon moves between the Earth and the Sun and blocks the sunlight reaching a certain part of the Earth. It sounds quite mundane, but in reality it is a glorious spectacle. The beauty of a total solar eclipse occurs as a result of a truly cosmic coincidence. The Sun is 400 times the diameter of the Moon, but the Moon is 400 times closer to the Earth, so viewed from the Earth, the Sun and Moon look as if they are the same size. When the orbits of the Sun and Moon intersect, the Moon perfectly covers the Sun, giving us the majesty of a total eclipse. The magic of a total eclipse is a truly cosmic coincidence and is a transitory effect. Although the position of the Moon seems stable, it is actually moving away from the Earth (at about the rate our fingernails grow). This means that in the past it was much closer and therefore appeared much larger in the sky, and in the distant future as it moves away it will appear to be much smaller and will not totally cover the sun. It is only in this epoch that the Moon's distance is just right to perfectly cover the Sun.

An interesting question is why do we not see total eclipses every month as the Sun, Moon and Earth align in their orbits. In the classic model of the solar system, we imagine each of the

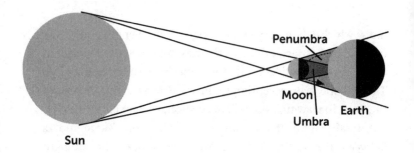

Penumbra: The Moon's faint outer shadow in the region partial solar eclipses are visible.
Umbra: The Moon's dark inner shadow. From this region a total eclipse will be seen.

orbits to be perfectly aligned with one another, all sitting within the same plane. This is not actually true, as small tilts in the alignment mean that we can seldom get the conjunction of all three bodies that is needed for a total eclipse.

I was fortunate to see a total eclipse in 1999, when I travelled to Le Havre in France, in an area where the total solar eclipse would be viewable. It is one of the most memorable events of my life. I was very lucky, as the day started cloudy but the skies cleared about 10 minutes before the event, so I could see it clearly. Wearing my solar eclipse glasses, I noticed the sunlight dimming to something like twilight, birds started to prepare to roost as the ambient light became dimmer, and finally there was the moment of totality when the Moon passes precisely between the Earth and the Sun. At this moment, you get a brilliant halo of sunshine visible around a dark Moon. It is also possible to see prominences (spikes of solar material) flowing off the Sun's surface. Totality lasted for a few minutes and as the Moon appeared to move on, I saw the brilliant diamond 'ring effect' when the first rays of sunshine shine past the Moon, then the usual daytime brightness returned. It was truly awe-inspiring and I remember having to sit down for a few minutes after the event to recover.

Any observations of the Sun aided or unaided should always be performed with suitable eye protection. Either solar spectacles for naked eye observations or the aforementioned 'H alpha' filter for telescopes and binoculars.

MOON GAZING

Observations of the Moon have no such protection limitations and it can be viewed in far from optimal conditions due to its brightness. Growing up in London, I think that I cut my metaphorical astronomical teeth by watching the Moon – I do see myself as a bit of a 'lunatic.'*

---

* The term lunatic originally meant a person who is under the influence of the Moon. This is because the Roman goddess of the Moon was called Luna and from the time of the Ancient Greeks until quite recently, many people believed that the full moon had a negative effect on the behaviour of vulnerable individuals and animals, causing them to go temporarily mad or act out of character – hence the term 'lunatic'. Fortunately, modern psychologists have disproved this theory, and there is no evidence that a full Moon has any affect on our behaviour.

When I say that the Moon is bright, it is important to note the vital role that the Sun plays in our observations of the solar system, as the Sun is the source of light that allows us to observe other objects. The Moon and the planets do not generate any visible light of their own, so to see them we need the sunlight that reflects off these objects.

If we look at the Moon, we can see some detail over its surface. Dark patches known as the 'Lunar Maria',* can be distinguished. These are areas of basalts that were formed billions of years ago when there was volcanic activity on the Moon. The lighter areas of the Moon, known as the 'Highlands', are older than the Moon's seas and are made of a rich mineral mix of aluminium and calcium. Many people in the past have combined these light and dark patches to form images. The most famous in Western culture is the 'Man in the Moon', but other cultures see different forms: for example, in China and the Americas, indigenous cultures called it 'Rabbit in the Moon'.

The Moon is an ideal candidate for children to observe. There is a great deal of mystery associated with it, and its phases are interesting to discuss. These are caused by how much the Moon's surface is illuminated by the Sun as it orbits** around the Earth every 28 days.

---

* Latin for seas or waterless seas.
** There are two ways to measure the duration of the Moon's orbit. The first is the sidereal month. This is the time the Moon takes to complete one full revolution around the Earth with respect to the background stars. However, because the Earth is constantly moving along its orbit around the Sun, the Moon must travel slightly more than 360° to get from one new Moon to the next. Thus, the synodic month, or lunar month, is longer than the sidereal month. A sidereal month lasts 27.322 days, while a synodic month lasts 29.531 days. People average these two values to get 28 days.

Sunlight

New

Waxing
Cresent

Waning
Cresent

1st Quarter

Earth

3rd Quarter

Waxing
Gibbous

Waning
Gibbous

Full

Lunar eclipses Just as there are solar eclipses, lunar eclipses occur too. A lunar eclipse happens when the Earth passes fully or partially between the Sun and the Moon, throwing the Moon into shadow. Because the Earth is much larger than the Moon, a total lunar eclipse often lasts a few hours. A total eclipse of the Moon is a rather wonderful, if slightly creepy sight, as the Moon turns a deep-red colour, often described as blood-red.

With the Moon in the shadow of the Earth, one would expect little sunlight to reach it as the Earth effectively blocks sunlight. However, sunlight can be refracted or bent by the Earth's atmosphere, allowing some sunlight to reach the Moon. This explains why we can see the Moon during a total lunar eclipse, but it does not explain why it goes blood-red. This reddish colouration occurs as a result of an atmospheric phenomenon called 'scattering' – particles in the atmosphere scatter shorter light wavelengths (blue light), and leave the longer wavelengths (red light) intact.

Most of us have seen this effect on a more local level when observing the setting Sun. Looking at the Sun around midday, when it sits high up in the sky there is less atmosphere scattering the light, so the Sun looks yellow. When the Sun is low on the horizon, the light travels through thicker amount of atmosphere and the blue light get scattered out leaving us to see the red light – which is the reason sunsets have that distinct red-orange hue. The blue light scattered by the atmosphere is what makes the sky appear to be blue.

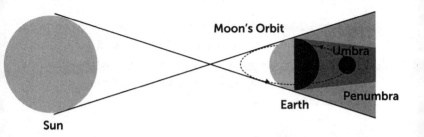

SHOOTING STARS

An enjoyable night's viewing is to observe the wonder of shooting stars. The term 'shooting star' is rather a misnomer because it does not involve stars at all but meteoroids. A meteoroid is a small rocky or metallic object travelling through space that varies in size from a grain of dust to an object about 1m (3ft) in diameter. Meteoroids originate from debris left behind by comets, and sit out in space a little like the slime trail left behind by a snail. A shooting star occurs when one of these objects interacts with the Earth's atmosphere. Due to its relative speed, a meteoroid entering the atmosphere heats up as a result of friction between itself and the particles in the atmosphere, similar to a spacecraft gaining re-entry. As the meteoroid reaches burning point, it emits a flash of light known as a meteor. Most meteoroids are small and completely burn up during this process, but if any part of a meteoroid survives and lands on Earth, the resulting object is known as a meteorite. Meteors often display different colours depending on their chemical composition.

METEOR SHOWERS

A meteor shower is an excellent time to see a large number of meteors. A meteor shower is an event in which many meteors are visible across the sky at the same time and apparently originate from the same point. During a meteor shower around 10 to 20 meteors can be seen every hour. Because certain comets have a regular orbit pattern, it is possible to predict when meteor showers will occur at different times of year. The showers are given names such as the Geminids (associated with the asteroid Phaeton) or the Leonids (associated with the comet Tempel-Tuttle), depending on which constellation the meteors seem to be originating from.

The best way to observe a meteor shower is unaided so that you can take in as much of the sky as possible. A shower usually reaches peak intensity over a period of two to three nights, so there is every chance you will have the opportunity to observe this spectacular cosmic firework display.

THE PLANETS

From shooting stars we move on to planets. Many of the planets in our solar system can be seen clearly with the naked eye, but it can be hard to distinguish them from stars. The trick I use to tell if I'm looking at a star or a planet is to see if the object being observed is twinkling. Many objects that we see in the night sky twinkle* due to pockets of turbulent air in the atmosphere (this visual blurring varies depending on the amount of turbulence there is on a given night and location). It's very similar to a heat haze on the road or disturbed air over a candle. Twinkling stars are considered by some to be beautiful or romantic, but astronomers will go a long way to avoid the conditions that cause this visual disturbance as it makes scientific measurements much harder and even astrophotography challenging. This is why astronomical observatories are often built in mountainous regions where the Earth's atmosphere is thinner and the air is drier, both of which minimise the amount of atmospheric turbulence. If you're not sure whether the body you're observing is a planet or a star, notice how much it twinkles. Planets, although much smaller than stars, are much, much closer to us. As a result they can be seen as a small disk through a

---

* The atmospheric depth is important for stargazing as stars that appear at the zenith of the sky are easier to see with less atmospheric disturbance because there is less atmosphere for the starlight to pass through. Stars at the zenith tend to twinkle less.

telescope. Stars, on the other hand, appear as pinpoints of light. Atmospheric turbulence will make a star twinkle, but will have far less of an effect on a planet, which will seem not to twinkle at all.

FINDING YOUR WAY ROUND THE SKY

As the planets of the solar system orbit around the Sun, they sit in different parts of the sky at different times of year. Sometimes a planet cannot be observed at night because it is in a position in its orbit where it can only be seen in the day and sits below the horizon at night. To see what is observable in the night sky at your location at a given point in time (there are many sources of information available). These days I mostly use a smartphone app. Using a smartphone's built-in GPS (Global Positioning System), it can map out the bodies that should be visible from your location right now. As you orientate the phone's screen, the objects that are viewable will change, giving you a map of the night sky in front of you literally at your fingertips. These are really useful for the beginner, and a wide range of apps exists that are often free to download.

There are also many guides on the Internet that can be looked up in advance or printed out. In the past I often used a planisphere, which is a type of star map. This, again, will show you what can be seen in the night skies in your location at different times of year.

SIX VISIBLE PLANETS

Being able to see the planets with just my eyes gives me a real visceral delight. Most of the planets in the solar system can be seen with the naked eye, and many of them have a distinctive hue that can help to identify them.

Venus This is one of the brightest planets in the night sky, partly because it is the closest planet to the Earth, and also partly thanks to the thick sulphurous cloud that surrounds it, which reflects up to 70 per cent of the Sun's rays. It's often called the Evening Star or the Morning Star, which gives you an idea of when to see it: just after sunset and just before sunrise.

Mars When visible in our skies, Mars is distinguished by its fiery, red-orange hue and minimal twinkling.

Jupiter and Saturn In a similar way to Mars and Venus, Jupiter and Saturn look like bright, non-twinkling stars. I usually refer to a star map to confirm their locations.

Mercury The same goes for Mercury, which is an elusive planet to spot due to its proximity to the Sun – you'll definitely need a star map to find this planet, and late spring is a good time to look for it in the northern hemisphere.

Neptune For those with very good eyesight and in optimal viewing conditions, it is possible to see Neptune unaided. It is not something that I've achieved yet, but I do hope to in the future.

STARS AND CONSTELLATIONS

From the planets we move on to stars and constellations. A constellation is a group of stars that early stargazers linked together to form an image (see page 72-3). I like to see constellations as a familiar 'word' in an otherwise foreign language – something that's recognisable among lots of noise. My heart still sings as I spot, for example, the familiar constellation of Orion over the winter skies of my home.

By their very nature all constellations are visible to the naked eye, so in this section I would like to mention a few that might be of special interest to the naked-eye observer. It is interesting to note that not all constellations are visible from all points of the Earth. If the Earth is fixed at the centre of the imaginary celestial sphere (see page 50), then our viewpoint in the northern hemisphere is different to that of the southern hemisphere. Due to the orientation of our planet, we have, to a certain extent, a fixed part of the heavens we can see. Hence the constellations of the southern hemisphere are quite different to those viewed in the north. There is some overlap, however: constellations that lie close to the ecliptic (the path of the Sun) which can be seen in both hemispheres. Also owing to the tilt of the Earth's axis, some constellations north and south of the ecliptic can be seen at different times of year in the northern or southern hemispheres. Orion, for instance, is a winter constellation in both hemispheres.

CONSTELLATIONS AND ASTRONOMICAL GEMS THAT ARE EASY TO SPOT

ORION

The constellation Orion, or the Hunter, is often one of the easier groupings of stars to spot, mainly due to the distinctive line of three bright stars that make up the figure's belt. Orion can be best seen in the northern hemisphere between November and February, and in the southern hemisphere it is best viewed in the summer months.

It may come as a surprise, but in the same way as some planets have very distinctive colours, some constellations have them too. This can be seen beautifully in the stars that make up the Orion constellation. Betelgeuse, on Orion's shoulder, is a red supergiant star (it's so massive, guestimates put it at anything between five and 30 times the mass of the Sun) approaching the end of its life. Easy to spot because of its orange-red colour, it is one of the brightest stars in the night sky. Compare this to the even brighter Rigel on Orion's left foot: a cool-blue supergiant that scientists predict will eventually explode into a supernova.

The stars that make up Orion's sword are also of great astronomical interest, although the middle object is actually not a star at all, but a nebula where new stars are being formed. It seems amazing that such an object can be seen with the naked eye.

THE PLOUGH

A very familiar asterism* is found in the constellation of Ursa Major, or the Great Bear. This can be seen all year round in the northern hemisphere. The seven brightest stars of Ursa Major make up the asterism called the Plough, also known as the Big Dipper. The Plough is fairly easy to spot and can be used as a navigation tool – a line drawn from the two stars that make the edge of the Plough, Merak and Dubhe, and extended by three times the length takes us to the Pole Star, Polaris.

POLARIS

In the northern hemisphere, Polaris, often referred to as the Pole Star or the North Star, sits directly above the North Pole, at the point of the Earth's axis of rotation, in the constellation of Ursa Minor, or the Little Bear. This means that Polaris never rises or sets, while all the other stars appear to move around it. This makes it an invaluable navigation tool, one that has been aiding the traveller for centuries to move close to true north.

THE PLEIADES

Another interesting constellation that can be viewed with the naked eye in both hemispheres consists of the seven fairly young stars that make up the Pleiades, or Seven Sisters. Visible

---

* A recognisable pattern of stars that is smaller than a constellation and may form part of one.

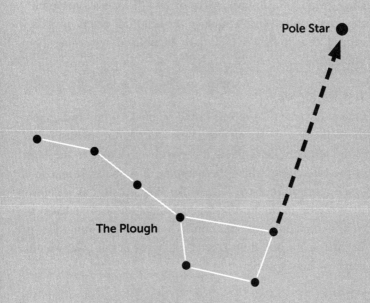

Pole Star

The Plough

in the northern hemisphere from late summer to early autumn, this cluster of stars is relatively close to the Earth; they look distinctively blue and have been compared to sapphires. In Greek mythology, the Pleiades were the daughters of Atlas, an early divine being known as a titan, and Pleione, a sea nymph – the legend is that Orion pursued the beautiful sisters until they begged Zeus to deliver them, and he obliged by turning them into stars.

THE MILKY WAY

One of my favourite objects in the night sky is our own galaxy, the Milky Way. Although it can be seen from both hemispheres, in the southern hemisphere it is possible to look into the heart of the Milky Way where the majority of stars reside and so is much more impressive. The Milky Way appears as a defuse, starry cloud in the northern hemisphere and can be tricky to view, so minimal light pollution is essential if you want to see it. Summer is the best time to catch a glimpse of this beauty in the northern hemisphere.

There are many more things to see in the night sky and it can be useful to keep a record of what you have been able to observe in form of an astronomical diary. There are lots of templates available for this and you can also include small sketches of what you have observed.

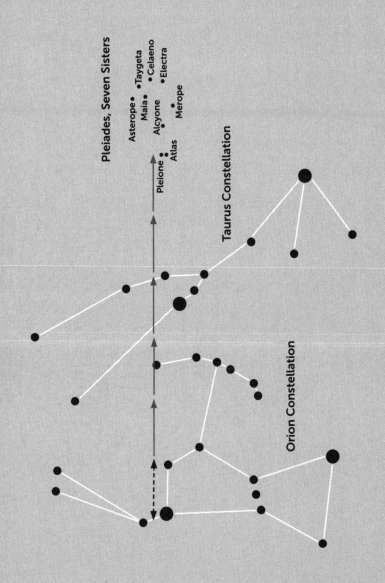

Pleiades, Seven Sisters

Asterope • • Taygeta
Maia • • Celaeno
Alcyone • • Electra
• Merope
Pleione • •
Atlas

Taurus Constellation

Orion Constellation

5

BINOCULARS
AND STARGAZING

Often when people get the astronomy bug the temptation is to go out and buy some great kit. Usually that kit takes the form of a telescope, but while you are finding your astronomical feet a good set of ordinary binoculars can be a wonderful starting point before making a major investment. They are easy to use (conversely, viewing objects through a telescope can take some practice), very quick to set up if the conditions are changing and usually much cheaper than a telescope. I would recommend them to anyone taking their next steps after naked-eye observing.

HOW BINOCULARS WORK

Binoculars and telescopes work on a similar principle: to gather as much light as possible and then display a magnified image on your retina.

At the front of the binoculars there is a lens called the objective lens. Its role is to gather the light from the object you want to observe and transmit this to an eyepiece at the other end of the binocular. The magnification of the binoculars depends on the shape of the objective lens, which is convex (dome-shaped) and magnifies images by bending light rays in towards the middle, thicker part of the lens, thereby making an object look bigger.

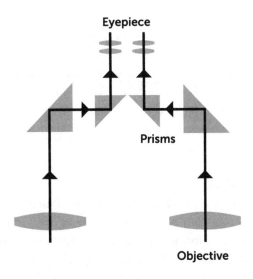

Eyepiece

Prisms

Objective

The focal length of an optical system is the ability of that system to converge or diverge a beam of light. For telescopes, the longer the focal length of the system, the greater the magnification that can be achieved. In a binocular system, a variation of the telescope, prisms (large wedge-shaped pieces of glass) are used to fold the light within the system, allowing a long focal length in a compact body. The prisms also invert the image to the correct orientation. Without them, anything observed would be upside down and back to front – not a problem for most astronomical observation but not so good for looking at, for example, nature or watching sports events.

After the light passes through the prisms, it passes through an eyepiece and into your eye. Further magnification can be achieved at the eyepiece. Many of us have a slightly different focus from one eye to the other. With some binoculars these small differences can be accommodated for by an adjustment at the eye piece. Usually labelled +/- so that a small level of focusing can be done.

MAGNIFICATION

The capabilities of a set of binoculars are described by a set of numbers. The first number states the magnification of the binoculars. So a binocular designated 7 x 50 means an object will look seven times closer than it would be if observed with the naked eye. For a good pair of astronomical binoculars, the magnification value should lie between 5 x and 12 x. Any higher than this and it will be hard to keep the binoculars steady enough to make decent observations, as with these levels of magnification the image will look as if it's dancing around and will be disappointingly blurry.

APERTURE SIZE

The second number given on a pair of binoculars indicates the size of the aperture: i.e. the objective lens, usually in millimetres. This number is critical if binoculars are to be used for astronomical purposes as it identifies how much light can enter the system. This is definitely a case where bigger is better.

In the case of field glasses used to observe nature, there is usually a lot of ambient light around, reflecting off the objects being viewed, so a small aperture is ample. However, in the case of astronomy the light levels from objects being observed are low and a larger aperture is required to gather more light. If we take the example of two sets of binoculars, one with 10 x 35 and the other with 10 x 50, although the second pair is only 15mm ($\frac{1}{2}$in) larger than the first, in terms of light-gathering power the second pair has twice the light-gathering capability of the first. This is because the area of the objective lens is twice that of the first. So a set of binoculars with a 50mm (2in) aperture will increase its light-gathering capabilities by around 70 times, meaning that much dimmer objects can be observed.

EXIT PUPIL

The exit pupil is the aperture that allows rays of light to reach your eyes. You can see the exit pupils in the eyepieces of the binoculars by the bright circles of light that appear when the aperture is fully illuminated. The size of the exit pupil is useful to look at when assessing binoculars. It can be calculated by dividing the diameter of the objective lens by the magnification. By matching the exit pupil to the size of your pupils you can increase viewing efficiency. After the age of 25, the average pupil size decreases from around 7mm to 5mm. If the exit pupil of the binoculars is too large, then some of the light gathered by the system is wasted, as it will not reach your retina.

FIELD OF VIEW

The magnification of the binoculars also affects the field of view that can be observed. The field of view is a measure of the amount of sky that can be seen through the binoculars – the greater the magnification of the binoculars, the smaller the field of view visible. For astronomical observing, a wide field of view is desirable as more sky can be seen, thus making viewing more pleasurable and enabling easier navigation across the sky.

EYE RELIEF

For individuals who wear glasses another important factor is eye relief. This is the distance between the eyepiece and your eyes while the image is in focus. A longer eye relief may make viewing easier for glasses-wearers.

AVERTED VISION

This is an observing technique where you look slightly away from an object through the binoculars (this also applies to telescope viewing; either by shifting your gaze to the left or right, depending on which is your dominant eye. Just as most people are right- or left-handed, we also have a dominant eye. To establish which eye is dominant, try this quick test. Hold your arms in front of you and use both hands to form a triangular-shaped viewing hole. Look at a nearby object through this hole, such as a doorknob. Now view the object with one eye closed and then the other – your dominant eye is the one that can see still see the object in the centre of the hole. With averted vision, you move in the direction of your dominant eye; so, for example, if you are right-eye dominant, shift your gaze slightly to the right. When you do this, the image falls on an area of your eye that is more sensitive to low light levels (see page 44-5), and is very helpful when viewing faint objects. Using averted vision can take some practise, but it's worth mastering this technique if you want to enjoy the finer details of the night sky.

COMFORTABLE VIEWING

Binoculars suitable for astronomy have a wider aperture than field glasses, so to observe happily for many hours it is often a good idea to be able to support them or your arms; a wall or fence is ideal. If these options are not available, a tripod can alleviate some of the strain. As well as supporting your muscles, having a support for your binoculars helps to steady the image while observing.

MOON THROUGH BINOCULARS

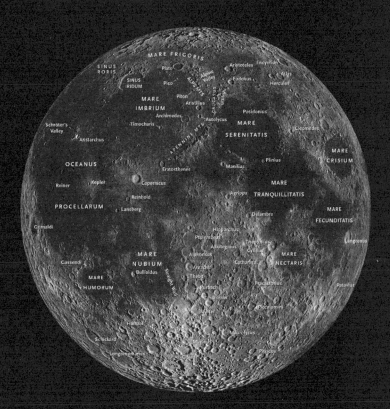

MARE FRICORIS

SINUS
RORIS
Plato
Aristoteles
Endymion
Atlas
SINUS
IRIDUM
Pico
Alpine
Valley
Eudoxus
Hercules

Piton
ALPS MTS

MARE
IMBRIUM
Aristillus
Posidonius

Schröter's
Valley
Archimedes
Autolycus
MARE
SERENITATIS
Cleomedes

Aristarchus
Timocharis
CAUCASUS MTS

OCEANUS
Eratosthenes
Manilius
Plinius
MARE
CRISIUM

Reiner
Kepler
Copernicus
APENNINE MTS
Agrippa
MARE
TRANQUILLITATIS

PROCELLARUM
Reinhold
Lansberg
Delambre
MARE
FECUNDITATIS

Grimaldi
Hipparchus
Ptolemaeus
Theophilus
Langrenus

Albategnius
Cyrillus
MARE
NECTARIS

Cassendi
MARE
NUBIUM
Alphonsus
Catharina

Bullialdus
Arzachel
Fracastorius
Petavius

MARE
HUMORUM
Thebit
Straight Wall
Purbach
Piccolomini

Werner
Walter

Hainzel
Maurolycus

Schickard
Maginus

Longomontanus
Pitiscus

Clavius

MOON LANDINGS THROUGH TELELSCOPE

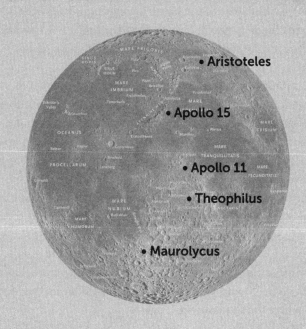

USING BINOCULARS

Viewing the night sky with binoculars opens up new discoveries, so let's look at some of my favourite objects and see what added delights binoculars will give us, investigating a few new bodies that are ideally suited to being observed with binoculars.

THE MOON

While it is possible to see some details of the Moon's surface with the naked eye, such as maria (lowland areas that show as dark patches) and terrae (lighter-coloured highland areas), with binoculars, suddenly craters can be seen in detail. The best time to observe the Moon is not when it is full. In this aspect, it is hard to see the fine points when they are directly illuminated from overhead. Better to observe the Moon when it's waxing or waning so that any craters are thrown into relief by shadows and you can see many more features. Focus on the area at the terminator line, which is the moving line that separates day and night on the Moon (this shows up very distinctly on the Moon), and where the shadows are longest.

JUPITER

The planet Jupiter is an ideal object to view with binoculars. With the naked eye Jupiter looks like a bright, slightly coloured non-twinkling star. With binoculars, Jupiter can be seen as a coloured oval-shaped disk, and four of its main moons, Io, Europa, Ganymede and Calisto, are also visible as bright specs in the night sky. If you observe these moons patiently for several hours you will see them moving around Jupiter as they orbit the planet. These are known as the Galilean satellites, as Galileo first discovered them around 1610, and are some of the largest objects in the solar system. Ganymede alone is larger than the planet Mercury.

ORION AND THE M42 NEBULA

Surprisingly, many distant astronomical objects beyond our galaxy are ideally viewed with binoculars as, with their wide field of view, extended objects such as galaxies, nebulae and star clusters can be observed with ease. The constellation Orion, for example, is impressive when viewed with the naked eye, but with a set of binoculars Orion has so much more to offer.

If you look below Orion's belt of three stars (Zeta, Epsilon and Delta), travel south down to Orion's sword to find another three objects, which to the naked eye look like a line of three fuzzy stars. Look at them through your binoculars and you'll see that these are not individual stars but actually three star clusters. In the middle of this group lies the famous M42 nebula, or Orion Nebula, which is one of the most photographed regions of the sky*. A nebula is an interstellar cloud of dust, hydrogen, helium and other gases. To view M42, it's actually best to look at it with averted vision. The amazing star facts

* The 'M' in M42 stands for Messier, the surname of the 18th century French astronomer Charles Messier. He observed and catalogued over 100 astronomical objects during his lifetime, and allocated a number to each of his identifications.

about the Orion Nebula is that it is around 14 light years across and sits 1,500 light years away from the Earth, a vast stellar nursery where stars and whole solar systems are born.

THE PLEIADES

The Pleiades (see page 72) has been a highlight in our skies since ancient times. It is mentioned numerous times in the Bible, and noted by virtually every other culture. It is located near Venus and has a distinctive dipper shape – can be seen with naked eye but looks even better with binoculars. The Pleiades is actually a small star cluster, with seven bright stars sitting in a nebula that contains around 3,000 stars. The Pleiades is one of brightest and closest star clusters in the sky and lies about 400 light years from the Earth. It is another birthplace for new stars.

ANDROMEDA GALAXY (M31)

Andromeda is a huge spiral galaxy containing up to 1 trillion stars that sits a whopping 2.5 million light years away. Even with this vast distance it is just visible with the naked eye as a fuzzy blob (Andromeda is the most distant object that can be seen unaided). As with the Orion Nebula, Andromeda is best viewed with averted vision. With binoculars, it is possible to see more detail in the glorious M31, and also one of its companion dwarf galaxies, M32. Interestingly, our galaxy and Andromeda are moving towards each other and are due to collide in approximately three billion years. If any of us are still around it should be a truly magnificent sight to see.

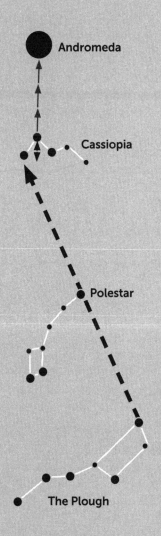

6
CHOOSING YOUR
FIRST TELESCOPE

A telescope can transform your astronomical viewing. With the unaided eye, the planet Saturn, for example, looks simply like a bright star. When viewed through binoculars, you can appreciate its ethereal golden colour, and notice the bulges around its edge. When viewed through a telescope, even one with a small aperture (60mm/$2\frac{1}{2}$in), you can see a distinctive disk shape and its famous rings are clearly visible, as well as its larger moons, such as Titan, Rhea and Iapetus, that orbit the planet.

When buying your first telescope, there are a lot of questions you need to ask in order to obtain the right system for what you wish to observe. One of the first questions is, what do you want to look at? Some people like to use their telescope to see the Moon and the planets of our solar system – objects that are relatively close and bright – in more detail. To do this, magnification is a key requirement and the size of the aperture is less critical. However, others would rather look at the multiplicity of deep-sky objects, such as stars, nebulae and galaxies. Many of these objects are large and extended in the night sky, so large magnification would not do them justice, as much of their substance would lie outside the field of view. These objects by their very nature are extremely far away, so to observe them we need to gather as much of the faint light as there is using a telescope with a large aperture. A wide aperture can be problematic, though, if the telescope becomes too large and unwieldy, which may make it awkward to carry and to store when not in use, so there's always a balance to be struck.

Of course there are telescopes that offer a reasonable balance between the two types of observation, and we will discuss how this can be achieved.

TYPES OF TELESCOPES

There are a number of different telescope designs available. The three main types are refracting telescopes, reflecting telescopes and catadioptric telescopes.

REFLECTING TELESCOPES

A reflecting telescope uses one or more curved concave mirrors to bend light to form an image. The advantage of reflecting telescopes lies in the fact that because light is merely bouncing off the surface of the mirror, the quality of the glass is less critical (although an accurate surface shape and finish and good thermal properties are required). This means larger apertures can be achieved, making a reflecting telescope ideal for viewing deep-sky objects and cheaper. Professional telescopes have become larger and larger, moving towards the reflecting telescope system. To make a large refracting telescope you need to have a piece of glass with good homogeneity and with no

TYPES OF TELESCOPE

REFRACTING

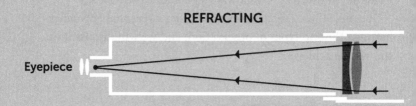

Eyepiece

REFLECTING

Eyepiece

CATADIOPTRIC

Eyepiece

inclusionsor distortions; getting large optics of this quality is very expensive if you want an 8m (26ft) lens. However, if using a mirror, the internal quality of the glass is far less critical, so it is possible to make larger mirrors with less quality which limits expense. This, in general, makes reflecting telescopes cheaper than refracting ones.

Advantages of reflecting telescopes
• Easy to use and assemble
• Excellent for viewing deep-sky objects
• Deliver clear, bright images
• Cheaper than refracting telescopes

Disadvantages of reflecting telescopes
• Not as good for viewing nearer planets and objects, although the focal ratio can help give you the best of both worlds
• As the tube is open, dust and dirt can gather on the mirror surfaces which will need to be cleaned more frequently
• Telescope mirror and tube can be larger, which makes them less portable

REFRACTING TELESCOPES

These use an objective glass lens with a large aperture. The lens bends light as it passes through the glass to form an image. The quality of the glass that the light passes through has to be very good so as not to cause aberrations (defects in the image quality), and this makes refracting telescopes expensive. However, there is a new resurgence in refracting telescopes manufactured by China. These are well made, with apertures of around 100-150mm (4-6in) and are good for looking at objects in our solar system.

Advantages of refracting telescopes
• Easy to use and reliable
• Excellent for viewing planets and objects in our solar system
• Can be smaller and lighter than their reflecting equivalent
• Usually similar set-up to terrestial telescope: i.e. looking through the end of a barrel

Disadvantages of refracting telescopes
• The aperture in reflecting telescopes aimed at amateurs is on the small side: typically 76-127mm (3-5in)
• Not as good for viewing deep-sky objects
• The need for good-quality glass can make them more expensive
• More likely to find cheap, low-quality versions in this telescope type

CATADIOPTRIC TELESCOPES

Sometimes also referred to as a compound telescope, a catadioptric telescope is a device that combines a mirror system with lenses to make a compact and relatively large aperture. The main downside to a catadioptric telescope is usually the cost, but if you can afford such a telescope, it is an excellent compromise for solar system and deep-sky observing, and is the most popular type available.

Advantages of catadioptric telescopes
• Offer great versatility, with option to view near objects as well as deep-sky ones
• You get the best of both refracting and reflecting telescopes
• Usually quite compact for their size, making them much easier to transport

Disadvantages of catadioptric telescopes
• Due to complex optics, these systems can be expensive

APERTURE SIZES

There are various numbers that describe the performance of an optical telescope and it is helpful to have an understanding of these before you buy.

Although a telescope can deliver magnification, one of its best functions is to gather more light. I always think of telescopes as light-gathering buckets, and the bigger the bucket (or aperture), the more light it gathers. The aperture of your telescope is the size of the main mirror or lens. The amount of light gathered by a mirror or lens increases as the square of its diameter (i.e. going from a 150mm/6in diameter to a 200mm/8in one provides nearly 80 per cent more light-gathering area, so about 80 per cent more light is gathered). Reflecting telescope apertures aimed at amateurs usually vary from around 100mm (4in) to 200mm (8in). They can go as large as 400mm (16in), but these telescopes are usually heavy and hard to handle. For observing objects in our solar system, a refractor telescope with a 100mm (4in) aperture or a reflector telescope with a 150mm (6in) aperture should do a good job. For observations of deep-sky objects, a reflector telescope with a 200mm (8in) aperture should be pretty ideal.

THE FOCAL RATIO OR F-NUMBER

The 'focal ratio' or f-number of a telescope is an important measure of its performance. This is the 'speed' of a telescope's optics, and is found by dividing the focal length of the optics by the aperture. The focal length of an optical system is the ability of that system to converge or diverge a beam of light. For telescopes, the longer the focal length of the system, the greater the magnification that can be achieved. For example, take a refractor telescope with a focal length of 1200mm (47in) and an aperture of 150mm (6in). Divide 1,200mm (47in) by 150mm (6in) to give an 'f-ratio' of 8.

The smaller the f-ratio (f4 to f5), the lower the magnification and the wider the field. Also the smaller the f-ratio for a telescope or lens the shorter the exposure time needed for a picture, hence systems with low f-ratios are called fast. Telescopes with low f-ratios are good for wide-field observing, but the image quality may be a little poorer than telescopes with a higher f-number. Slow f11 to f15 ratios are usually better suited to observing the Moon and planets in our solar system, slow here implies that a longer exposure time is needed compared to the smaller f-ratio systems. An f-ratio of 8 is a very good compromise if you want a telescope for viewing both our solar system and for deep-sky observations.

EYEPIECES

As well as the configuration of the telescope, further levels of magnification can be achieved via the use of different eyepieces. Many people will buy a small range of eyepieces to go with their telescope so that they can select the best eyepiece for the object they are viewing. Don't be tempted to buy eyepieces with very large magnifications, as these give a small field of view, and any vibrations picked up by the telescope will mean that the image bounces around uncontrollably.

The eyepiece supplied with a telescope may not be the best quality. The easiest way to upgrade the quality of your overall system will be to get a better eyepiece.

Once we have our main telescope system the best way to change the magnification is by changing eyepieces. Your basic repetoire will need three eyepieces. Firstly, you will need a high magnification eyepiece for observing detail on the Moon and planets. Secondly, you are likely to want a medium-magnification eyepiece for brighter deep-sky objects (star clusters, nebulae and galaxies) or when you need to check for more detail in these objects. Finally, you will need a low-magnification eyepiece that you will use mostly for locating objects. It is always better to do this with low magnification and a wild field of view.

If you wear glasses for correction of long- or short-sightedness then you should not need to wear glasses when observing, as a small adjustment will get you in focus. However, if you suffer from an astigmatism then, your glasses will be needed, but look for eyepieces that have a rubberised gasket so that your glasses do not get damaged.

Another useful addition to the collection is a Barlow lens. This optical device sits in your telescope ahead of the eyepiece and can double or triple your magnification. The insertion of a Barlow lens will, however produce a less-bright image.

TYPES OF MOUNTS

Once you have your ideal telescope, you need to mount it securely. A robust mount is critical for optimal viewing. If your telescope mount is not stable, then objects in your field of view will appear to move about as you view them and vibrations may prevent you from making quality observations.

There are two main types of telescope tripods: the first is a simple up-down, left-right mounting system called an alt-azimuth mount; the second, which is more sophisticated and makes tracking objects across the night sky easier, is called an equatorial mount.

THE ALT-AZIMUTH MOUNT

A fixed altitude mount, the alt-azimuth mount has an altitude axis that allows the telescope to be moved in an up-and-down direction, and an azimuth axis that allows the telescope to move from left to right. It is a simple system and is best used for making terrestrial observations as it is aligned with the horizon. However, this system makes it very difficult to track stars, because as the Earth rotates, you have to move both up and sideways to keep the object in the field of view. Without a computerised mount, there is no easy way to do this.

THE EQUATORIAL MOUNT

An equatorial mount makes viewing the night sky easier as it has a declination (DEC) axis that allows the scope to be moved north and south (not dissimilar to the altitude axis), and a right ascension (RA), or polar, axis, allowing the scope to be moved east and west (similar to the azimuth axis). The benefit of this design is that the polar axis of an equatorially mounted scope is aligned to the north (or south, depending on which hemisphere) celestial pole, and is therefore also aligned with the Earth's axis of rotation. This is the point in the sky around which all the other stars appear to rotate. This means that if you are observing a star, to track its progress across the sky you follow it just by using the polar axis rather than using both axes, which is the case with the alt-azimuth mount. For this reason many telescope systems opt for the equatorial mount for astronomical observing and imaging.

DIGITAL TRACKING

As a child, having made my own telescope I set about making a digital tracking system for it. My idea was to put the latitude and longitude of my location into a computer, set the telescope looking up at the zenith and then track other objects in the sky from my known location on the celestial sphere above. This was a labour of love as I input the details of thousands of stars into the system. Fortunately, digital tracking systems have come a long way since then. They are now small, accurate and built in to the telescope system. For anyone considering serious astrophotography of faint objects, a tracking system is a must as it allows you to get very long exposure while keeping the object fixed in the same position through the telescope. For others it may be a useful tool to help navigate across the sky.

Most modern telescopes with motorised drives have a 'go-to' mount. This allows you to select a target object and then software and drivers on the telescope mount will move the telescope to this location of the sky and track the object. The software in the mount will then move the telescope to this location of the sky and track the object until told to move elsewhere.

The information above is meant to be a general guide to what to look out for, and it's essential to get more detailed advice if you intend to make a significant investment. Many astronomical societies are happy to speak to people who are thinking of buying a telescope and to share their personal experiences. Better still, attend a meeting of your local astronomy club where you can try out some of the different types of telescope and ask lots of questions to help you work out what might be the best system for your personal use.

7
SETTING UP
YOUR TELESCOPE
AND WHAT TO
SEE WITH IT

You've done your research and chosen the telescope that's right for you – what follows is a rough guide to setting up your telescope and mount, which can in no way replace the detailed instructions that you will receive with your telescope, but may be useful for guidance.

When your telescope arrives, it makes a lot of sense to set it up in the daytime and I also usually start indoors so that I can keep track of all the bits. Your telescope and mount will arrive in parts, so lay everything out first and check that all the items listed in the instructions are there. While it may be stating the obvious, read the instructions!

THE TELESCOPE MOUNT

The telescope mounting (usually a tripod) is a critical component of your observing experience. Once you have assembled your mount, make sure that all the screws and wing nuts are tight. It is much easier to do this in daylight rather than in the dark with a torch. At this stage also check that your telescope support mechanism is level. Many telescopes come with a spirit level built in, but if this is not the case and you don't have one at home, hang a weight on the end of a string to ensure that your telescope mount is level.

Select a suitable spot to set up your telescope – for example, the back garden (this works for me, as the house blocks out light pollution from the streetlights). The mount needs to be placed on a sound, even surface to minimise wobble. Testing out a few locations may be useful. Mount your telescope onto its support and ensure all of the necessary bolts, screws and straps are in place and secure.

SETTING UP AN ALT-AZIMUTH MOUNT

For an alt-azimuth mounting system little adjustment is needed, but tweak the locking mechanisms so that you can move the telescope freely both up and down and left and right while ensuring that the telescope stays in place when it is not being touched.

SETTING UP AN EQUATORIAL MOUNT

An equatorial mount takes much more effort to align, but should be easier to use in the long run. An equatorial mount is aligned to the polar axis so that it is parallel to the Earth's axis. To make a correct alignment you need to find the latitude of your location and adjust the polar axis of your telescope to the same number of degrees from the horizontal. Many telescopes have a latitude scale printed on the mounting. This can be used as a good approximation, but should be confirmed by looking at the Pole Star. To do this, position the Pole Star, Polaris, in the centre of your field of view. Once this is done, tighten the adjustment, as this alignment will not change unless you take the telescope to a significant change in latitude.

The next step is very important, which is to balance your telescope with its counterweights. A telescope is a heavy object, so adding weights to counterbalance the weight of the telescope will ensure smooth observations about the axis of the mount. If this exercise is not done correctly, the telescope will move in an uncontrolled manner and tracking objects will be very difficult.

Your telescope will need to be balanced in both the RA and the DEC axes. To do this you need to set up your telescope as if you are about to do a night's observing, so put a low-power eyepiece into your telescope (this ensures you are counterbalancing the right sort of weight). The RA axis sits between the telescope and the counterweight. While supporting your telescope, loosen the RA axis so that the telescope can move freely, and then carefully loosen the counterweights on their shaft so that they are fairly free to move with a little force. Move the counterweights until they balance the weight of the telescope – this means that it should neither drop nor rotate when unsupported. When this is achieved, lock the counterweights into position.

For the DEC axis, you need to balance your telescope front to back. Equatorial mount telescopes usually have some sort of cradle holding the telescope. Others may have a slide mechanism. One of mine has a dovetail tongue-in-groove system that allows me to move the telescope forwards and backwards relative to the support mechanism. Release the bolts that keep this mechanism in place and move the telescope forwards and backwards until it feels as if it is in a balanced position. When this is done, retighten the bolts to lock the telescope in position.

THE FINDER SCOPE

If your telescope has a finder scope, this needs to be aligned to your telescope's line of sight. This small, low-powered pointing device, which sits on the main telescope, has a wide field of view and some kind of mark – usually a red dot or a crosshair – in the centre that indicates what the main telescope is pointing at. This is a very useful piece of kit to find your way round in the night sky, as trying to get your bearings with just your telescope – can be hard work – and frustrating, due to the telescope's higher magnification and narrower field of view. Even if the main telescope is misaligned by a small amount, it's very hard to tell where you are in the night sky. With a finder scope, you can easily find the object that you are searching for, set it to the crosshairs or red dot and it should be in the field of view of your main telescope.

Aligning your telescope and the finder scope is best done during the day, with the final adjustment made at night. Insert the lowest-power eyepiece into the telescope, then aim the telescope at a distinctive object around 30m (100ft) away (for example, the television antenna on a house or a distinctive cluster of leaves on a tree). Centre this object in the middle of the telescope's field of view, and then lock down the telescope's axis. Now find the same object with your finder scope and adjust the scope until the object sits at its crosshairs. Check that the object is still sitting at the centre of your main telescope's field of view. When it sits at the centre of both scopes, tighten all the screws of the finder scope. To finish the alignment procedure, run through the whole process again but at night, using a bright star as a target. Then do the process again, but this time using a higher-power eyepiece.

THE NIGHT SKY THROUGH THE TELESCOPE

So you finally have your new telescope and it is set up and ready for action. What do you look at? Here are some useful observations for beginners to cut their astronomical teeth on, objects that are easy to find and wondrous to observe through a telescope and that cover a diverse range of objects both near and far.

THE MOON

The Moon is the perfect celestial object on which to hone your telescope-viewing skills. Despite our familiarity with this satellite of the Earth, on the wow-factor front the Moon has still got it. To see as much detail as possible, it's best to view the Moon when it's not full. Using a magnification of around 60 times, the Moon should just about fill your field of view and you can start to observe details you would not have seen before (don't go for too a high magnification to start with; you want a wide field of view). If you catch the lunar bug and regularly observe the Moon, obtaining a good lunar map can do nothing but enhance your observations.

Here is a list of features to observe:

Maria, Craters, Ejecta plumes, Rilles, Mountains, Domes *

JUPITER

Moving further out into the solar system, our next hot observation is Jupiter (see page 27), which can only be viewed at certain times of the year, so check to see if it is visible. With as little as x 40 magnification, Jupiter will appear as big as the full Moon in the night sky. You should also be able to see Jupiter's four largest moons – Ganymede, Europa, Io and Callisto –

---

* **Maria** These are large dark areas on the surface of the Moon that are mainly basalt (exposed volcanic rock) that were formed by ancient volcanic eruptions. They originally got their name as people thought that they were oceans on the Moon's surface. **Craters** There are literally hundreds of craters over the Moon's surface. These vary in size but were generally formed by debris impacts occurring over billions of years ago. As the Moon has no atmosphere it has no protection from passing debris and once a crater is formed there is no weather to erode it away. Only other impacts occurring in similar locations will remove evidence of earlier impacts. The more prominent craters visible on the Moon's surface. Maurolycus is a large crater some 114km, (71 miles) across and 4,730m (15,500 feet) deep. It is located in the central region about a third of the way up from the south pole. There are a number of other smaller impacts over this craters' surface. Theophilus is a spectacular crater near the centre of the Moon's disc it is 100km (62 miles) wide. The crater's walls rise 1,200m (3,940 feet) above it's surroundings, it has a large central peak that lies below its rim. Aristoteles is another large centrally located crater near the Moon's north pole about ⅕ of the way down, it is 87km (54 miles) across. **Ejecta plumes** This is the remnants of debris expelled in a stellar explosion. **Rilles** These channel-like grooves on the Moon's surface are many miles wide and hundreds of miles long. **Mountains** Although there is no tectonic activity on the Moon, there are still a number of mountains and mountain ranges over the Moon's surface. These are formed by the result of impacts by asteroids in the past. When these impacts occurred they formed the Lunar Maria and left their rims to form lunar mountains. There are a number of these mountains that can be seen through a telescope. **Domes** Lunar domes were typically formed by erupting lava which slowly cooled down. The resulting lunar domes are circular features with a gentle slope which are usually a few hundred meters high.

with ease. Jupiter itself reveals a distinctive disk with coloured bands (known as belts) over the surface. With reasonable magnification, you may be able to just make out the Great Red Spot (also known as 'The Eye of Jupiter') in the atmosphere of the planet. This red spot (actually, it looks like more of an pink-orange colour), the result of Jupiter's tumultuous weather system, is a high-pressure, anti-cyclonic storm three times the size of Earth, and was discovered in 1664 by the English physicist Robert Hooke (1635–1703). It is still going strong over 400 years later!

ORION

By far one of my favourite constellations is the easily recognisable constellation of Orion, and another area of the night sky that is good to observe aided or unaided. With a reasonable-sized telescope, Orion can give up even more delights. Focus on M42 (see page 87), the Orion Nebula and a birthplace of stars. Hot new stars within nebula light up some of the interstellar gases, which emit light in characteristic bands. With a telescope you can observe some of these amazing colours in tints of green, red and brown. M42 is the nearest region of rapid star formation and sits 1,340 light years away from us. Within the Orion Nebula is a cluster of four stars that form a trapezoid shape known as the Trapezium. The Trapezium contains a number of young stars that formed out of its parent nebula. With a good-sized telescope six stars can be resolved. Galileo first discovered three of the stars around 1617, with the other stars being detected later.

GALAXIES

Galaxies are good targets for a new telescope. An old favourite that is also popular for naked-eye observation and when using binoculars is the Andromeda Galaxy (see page 88) and a telescope only brings out more of its glorious detail. If you are looking for something new, try tracking down the Triangulum Galaxy, otherwise known as M33, which lies to the southeast of Andromeda near the arrow-like asterism called the Triangulum. One of my favourite galaxies to view with a telescope is the very well-photographed M51, or Whirlpool Galaxy. Like the Milky Way, M51 is a spiral galaxy, and thanks to its orientation it is possible to admire its wondrous spiral arms. M51 is in the process of colliding with a smaller galaxy in its vicinity and this is thought to be generating more star formations in this region.

Key objects to view with a telescope
Hercules Global Cluster (M13)
Crab Nebula (M1)
Double Cluster (NGC 869 and NGC 884)
Ring Nebula (M57)
Lagoon Nebula (M8)
Pinwheel Galaxy (M101)

ADDITIONAL THINGS TO SEE

As well as stars, meteors and planets there are other things to see in the night sky with the naked eye. Try spotting some of the things listed below.

Satellites: Surprisingly enough it is possible to see these artificial objects from terra firms. This is because their solar panels reflect some of the Sun's light making them very bright in the sky. A satellite looks like a reasonably bright star travelling across the heavens at a fair clip. If you see coloured flashing lights, that is an aeroplane. A satellite emits a steady light and will be seen travelling a long distance till it disappears into the Earth's sun shadow.

Aurora: Visible from the extreme latitudes of Earth in the northern and southern hemispheres, this is the interaction of charged particles from the Sun interacting with the Earth's magnetic field. Their intensity depends on the levels of solar activity, and the colouring of the aurora is dependent on the chemical nature of the particles streaming through space.

Comets: In the past comets were seen as the harbingers of doom, but these heavenly bodies can be appreciated with the naked eye and binoculars. Comets are the debris left behind from the formation of the solar system. They are usually described as dirty snowballs that orbit the sun. Some can be clearly spotted from

Earth with the naked eye due to their distinctive tails which can streak a long way across the sky. Looked at with binoculars, more details can be seen, such as a central nucleus and the coma: a cloud of material around the nucleus.

If you do have access to a telescope here are some additional gems you may want the check out:

Star clusters: If you have a set of reasonable binoculars available other items to check out, are star clusters. Star clusters or star clouds are tight groupings of stars. These were made to be looked at with binoculars, as many have a large angular extent so are hard appreciate with a telescope. There are a number of them to appreciate, such as the aforementioned Pleiades and the Beehive, but one of my favourites is the Double Cluster. Two star clusters in close proximity to each other can be spotted in the constellation of Perseus.

Hercules Global Cluster: This globular cluster is a spherical collection of stars that orbits a galaxy as a satellite. The Hercules Globular Cluster was discover by Edmond Halley in 1714. It is made up of over a hundred thousand stars and it is an ideal target for a small telescope.

NEBULAE

A nebulae is a cloud of dust and gas found in space. They are a nursery for star formations and also mark the death of stars. The Crab Nebula sits north of the constellation of Orion. It is the remnants of a star that went supernova in 1054. It is possible to see the wisps of gas and dust left behind by this cataclysmic explosion as on oval splodge through a small telescope.

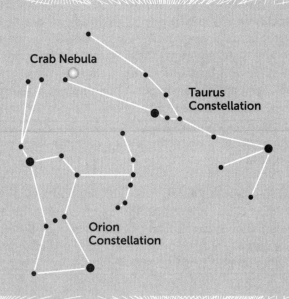

Crab Nebula

Taurus Constellation

Orion Constellation

PINWHEEL GALAXY

I have already mentioned the joy of observing the Andromeda Galaxy, our closest neighbour in space. Well, there are many more out there to appreciate. One of my favourites is the Pinwheel Galaxy. To me it looks as a galaxy should – spiral arms winding outward and it is relatively easy to find sitting a little north of the asterism the Plough (see page 89).

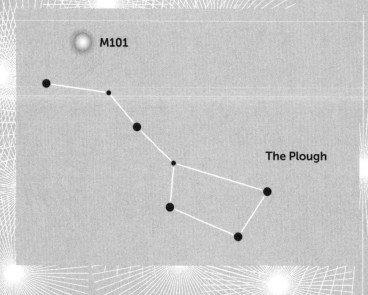

M101

The Plough

8
SIMPLE
ASTROPHOTOGRAPHY

Astrophotography is an area of photography that specialises in taking images of objects in the night sky. This activity may seem a little daunting at first to the amateur, but there are many benefits in taking the plunge. It's possible to take wonderful pictures at every level of involvement, from using a regular point-and-shoot camera with no other equipment, to using a telescope with a smartphone. One of the main benefits of astrophotography is the amount of extra light that can be gathered by this process. When we look through a telescope or binoculars, we are using the optics of the device to gather more light and feeding this into our eyes. When we photograph objects, longer exposures and increased sensitivity mean that even more light can be obtained, which allows us to discern more detail. Photographs can also be greatly enhanced using post-processing techniques.

ASTROPHOTOGRAPHY BASICS

You can take rewarding pictures of the night sky with anything from a compact digital camera (as long as it has manual settings too), a sophisticated DSLR (digital single-lens reflex) camera or an old-fashioned manual SLR (single-lens reflex) camera to a smartphone combined with a telescope. Whatever device you're using, the following covers the photography basics that you need to know in order to take a night sky image.

Most celestial objects are quite faint (with the exception of the Moon). As a result we need to gather as much light as possible in order to see as much detail as possible in the objects we're photographing. In photography there are three main ways to increase the light passing through the camera: exposure time, aperture size and detector sensitivity.

Exposure time and shutter speed Exposure time is generally governed by shutter speed: the longer the shutter is open, the more light that can enter and hit the detector at the back of the camera (film or CCD). In SLR cameras the shutter speed usually varies from $\frac{1}{1000}$th of a second to a few seconds. You will need a setting of between 10 and 30 seconds to take an effective night sky picture. With compact digital cameras, the night-mode setting offers a slow shutter speed for photographing at low light levels. It may also feature a firework setting, which may also be suitable for night photography. For smartphones, there are a number of apps available that effectively slow the shutter speed of your phone's camera. A useful tip if you use a digital camera – make sure you bring at least one extra set of batteries with you, as long exposures use up a lot of the battery's energy, and you will need to replace them regularly during an evening's session.

Many cameras also have a bulb setting for long-exposure shots. This enables the camera shutter to stay open for as long as you press the shutter-release button. An inexpensive cable release with a lock feature that plugs into your camera is useful for the bulb setting. This allows you to depress the shutter-release button remotely without actually touching the camera, and therefore avoids camera shake, which is important when you want to take sharp pictures in low light conditions. Alternatively, you can use the camera's timer.

Aperture setting The aperture of a camera controls the amount of light passing through the camera lens. It is set by a diaphragm that sits close to the front of the camera and varies in size, depending on the aperture setting. The aperture setting is governed by the f-stop value, which appears as series of numbers, usually f2.8, f4, f5.6, and f8, but more sophisticated DSLR cameras may offer a wider range of f-stop values. The smaller the number, the wider the aperture, so for most astrophotography f2.8 is the desired setting.

ISO speed This determines your camera's level of sensitivity to incoming light. Most digital cameras that have a manual mode offer ISO settings of 100, 200, 400 and 800, although again more sophisticated cameras will offer a wider range. The lower the ISO number, the more light that's required. So when filming in daylight, where there is plenty of light, an ISO of 100 or 200 is a suitable speed. For astrophotography, where the light levels are very low, an ISO setting of 400 or even 800 is desirable. For cameras that use film, you will therefore need film of ISO 400 or 800 for night sky photography. However, there is a penalty for this increased sensitivity, which is that higher ISO speeds will result in grainier images (often referred to as 'image noise'), although for smaller-image prints image noise is unlikely to be that noticeable.

Infinity focus To obtain a good image of the night sky, the camera focus needs to be set to infinity (if it has this setting). Infinity focus enables the camera to focus on objects that are essentially infinitely far away. On many point-and-shoot cameras this is indicated by the 'mountain' setting. For more sophisticated digital cameras, you will need to turn the autofocus off and set the camera manually to focus at infinity. It can be useful to try this out during the day when you can see faraway objects.

Camera tripod Another critical requirement for night photography is a stable position for your camera for the 30 seconds or so it will take to shoot the object. Ideally, you should use a sturdy tripod for this, but if you don't have one, a conveniently located garden wall would do. Having the camera stable for the duration of the exposure is essential. There are also tripods available for smartphones.

Experimenting with your photographs is a good idea; for example, try a few different exposures, and maybe even decrease the setting if you want to get a large depth of field (like a skyscape with earthbound structures). Keep notes and see what works best for you and your camera set-up: what is the best exposure, for example? Long exposures mean more light, but if they are too long you may get star trails (see below) where you don't want them. The joy of digital photography is that even if only one or two pictures have worked, throwing the others away has cost you nothing more than a little time.

READY, STEADY, ACTION

The optimal conditions for photographing the night sky are a clear, cloudless evening and minimal light pollution – that includes light pollution from the Moon, so choose a time before the Moon has risen or after it has set. If you want to photograph the Moon, choose a phase away from the full Moon so that the amazing details of its craters are thrown into relief by magnificent shadows.

Star trails One of the simplest images to obtain if you have a camera with a bit more control and clear skies is that of classic star trails. These photographs show the light trail of stars as they appear to wheel across the night sky in beautiful curved arcs. For star-trail images, the only equipment you need is a digital camera with manual modes, including a bulb setting, a cable release with a lock feature and a sturdy mount. Make sure you have a high-capacity memory card and check that there's lots of space left on it (also bring a spare, just in case). Back home, you will also need software for your computer to stack your images. While it's important to keep strong sources of light pollution to a minimum, with star-trail images a little light can enhance a picture by adding some terrestrial interest to your image, although too much light can obscure the stars you are trying to capture.

Once you have a found a suitable location, set up your tripod and mount your camera. Choose a patch of sky and set the focus to infinity. Although you can take a single shot of star trails, this involves a long exposure time (for example, up to 30 minutes), which can result in a lot of visual noise (see above). A more satisfying result can be obtained by taking lots of photographs about 30 seconds apart for the duration of 20 minutes or so, then using computer software to 'stack' (or merge) the images afterwards. As there are so many different types of digital cameras available, refer to your camera instruction booklet to set up the camera so that it will take an image every 30 seconds on continuous shoot mode – once you've pressed down the cable release and locked it, the camera will shoot an image every 30 seconds until you stop it. Take some test shots to make sure you are happy with the results, and make any necessary adjustments before leaving the camera to take multiple images.

The smartphone-telescope combination If you want to take your astrophotography to the next level, you can introduce a telescope to the proceedings. It may come as a surprise, but the Moon and even some of the planets can be photographed with a smartphone camera or a standard camera held up to the telescope's eyepiece. Focus the telescope on the Moon and then move the camera of the smartphone towards the eyepiece until a focused image can be seen on the screen – with such a bright object there is no need to worry about exposure time, so when it is in the right position just click and take the picture.

Piggybacking If you have a telescope with motorised tracking and you would like to capture fainter objects, a technique called piggybacking may be the solution for you. Fainter objects require long exposure – often of a few minutes and sometimes much longer than that (anything from 30 minutes to an hour and a half). The problem is, photographing anything in a stationary position for longer than 20 to 30 seconds or so results in star trails. With piggybacking, you use your telescope motorised tracking to keep the object that you are photographing stationary in the field of view by bolting the camera to the barrel of the telescope. You don't need any special lenses – a standard 35mm lens that you use for daytime photography is fine – to capture the beauty of, for example, the Milky Way in amazing detail and dazzling colour.

The camera-telescope combination The ultimate solution for astrophotography is to use the light-gathering power and magnification of a telescope with a motorised tracking system coupled directly to the camera. The telescope just acts like a giant lens to the camera. To do this, a two-piece T-mount is needed. The first piece, the T-ring, screws into the thread that the lens attaches to. For cameras without a removable lens, the T-ring attaches to the front of the lens where the thread for filters sits. The second piece is the T-adapter; this fits onto the telescope – many eyepiece holders have a thread on them that the T-adapter can attach to. This will screw onto the T-ring, linking the camera with the telescope. Once the camera is in place, the telescope will need to be rebalanced due to the camera weight. With this configuration, you can obtain some amazing long-exposure images, although you'll need to experiment with exposure times, aperture sizes and detector sensitivity to achieve the results you're looking for.

POST-PROCESSING

If you're using film in your camera and printing your own images, a certain amount of enhancement can be done through the developing and printing process. If you're using digital photography, then do investigate the huge range of image-processing software enhancement techniques that are being adopted by amateurs today.

To find out more detail about all of these techniques, look on the web, even better – I would highly recommend going along to your local astronomical society where other like-minded characters will be tackling similar challenges and may have some handy tips.

9
PROFESSIONAL TELESCOPES

As mentioned in Chapter 1, to my mind astronomy is the study of everything that is not of Earth. But this poses a problem, and it has all to do with the nature of space. When we are observing the universe on the ground, the thermal currents in the Earth's atmosphere distort any light rays coming from the night sky, making stars appear to twinkle and dance (see page 67). As we travel higher and higher above sea level, at greater altitudes the air gets thinner and thinner (so there is less atmospheric distortion), and the stars twinkle less. This is why many of the world's largest telescopes are located on mountainsides, where the amount of turbulent atmosphere above them is minimal. With the Hubble Space Telescope (see page 138) scientists went even further by launching a telescope into orbit in 1990 about 550km (342 miles) above Earth, at a distance where 90 per cent of the atmosphere is considered to be below us. At this height objects are virtually in the vacuum of space where observations should be much easier, although the vacuum can be challenging in itself. While a vacuum is a poor medium for transmitting most energetic waves, electromagnetic radiation travels well in it.

THE ELECTROMAGNETIC SPECTRUM

The electromagnetic spectrum is the entire range of electromagnetic radiation (EM radiation, or EMR), which is made up of different forms of electromagnetic energy waves (see diagram 132-3). All these waves travel at the speed of light, but they oscillate (or vibrate) at different frequencies. The building blocks of EM radiation are photons, which are minute particles of light that have no mass. Due to their nature, each of these forms of EM radiation can travel through a vacuum and so can therefore travel through space and propagate over vast distances. When we undertake the act of astronomy, amateur and professional alike, we are interacting with EM radiation that has passed through the vacuum of space to be eventually detected by ourselves. By analysing EM radiation, we can gain a better understanding of what is happening beyond the Earth.

Take a packet of EM radiation energy: a photon that has been generated in the heart of a distant star and has escaped from its surface. This photon, which may have travelled billions of miles through the vacuum of space, if passing in the right direction, will interact with the Earth. One of the other influences of how that photon will fare as it approaches the Earth is very much dependent on the Earth's atmosphere. This is because only some forms of EM radiation can pass through our atmosphere.

We can tell that visible light can pass through the atmosphere because that is what we can detect with our eyes and we can

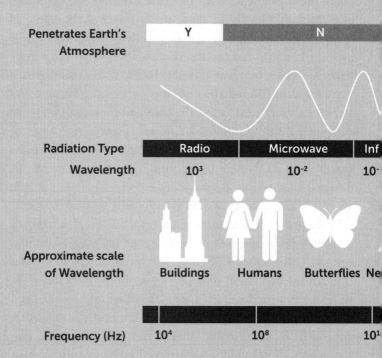

Penetrates Earth's Atmosphere

Y	N

Radiation Type	Radio	Microwave	Inf
Wavelength	10^3	10^{-2}	10^-

Approximate scale of Wavelength

Buildings	Humans	Butterflies	Ne

| **Frequency (Hz)** | 10^4 | 10^8 | 10^1 |

THE ELECTROMAGNETIC SPECTRUM

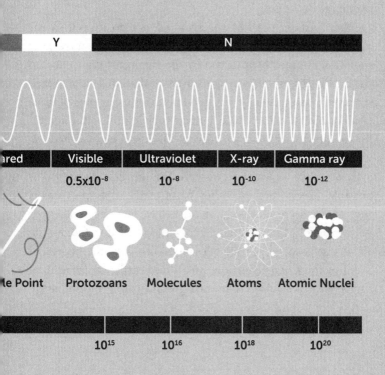

| | Y | | N | | | |

| | Visible | Ultraviolet | X-ray | Gamma ray |
| ared | 0.5×10^{-8} | 10^{-8} | 10^{-10} | 10^{-12} |

| le Point | Protozoans | Molecules | Atoms | Atomic Nuclei |

| | 10^{15} | 10^{16} | 10^{18} | 10^{20} |

observe the light of distant stars with no difficulty, so visible light passes through the atmosphere virtually unattenuated. Radio waves are similar to visible light. We bounce these waves across the Earth and use them as our main means of communication over large distances.

Other parts of the EM spectrum do not do so well in our atmosphere. X-rays, gamma rays and a lot of ultraviolet radiation are mostly absorbed by the chemicals that make up the Earth's atmosphere, so much of this radiation does not reach sea level. This is actually very good news for us, as each of these types of radiation can cause damage to living cells. In humans, for example, they generally disrupt DNA and cause cancer, so we're very lucky to have an atmosphere that protects us.

However, if we want to observe the universe as widely as possible, access to these other parts of the EM spectrum is very useful. To reach them, scientists have launched telescopes with special EM radiation detectors into space, to sit above the Earth's atmosphere. Space telescopes also have the benefit of avoiding the visual distortions mentioned earlier, so can achieve much better resolution than their ground-based counterparts.

GROUND-BASED AND SPACE-BASED TELESCOPES

Professional telescopes fall into two distinct categories of telescope: ground-based and space-based systems. I have been very lucky to work on both in my career, and there are amazing aspects to both types. To be part of a team working on a system that will one day be launched into space and help us gain knowledge of our place in the universe is a true honour. At the same time, while working on ground-based systems I have been filled with immense joy as the Sun sets and an observatory structure the size of a cathedral is opened for a night's observing – truly inspiring.

A few decades ago space-based systems were generally considered superior to ground-based telescopes. This was partly due to their ability to detect radiation that would otherwise be filtered out by the Earth's atmosphere and partly due to their location above the atmosphere, free from any turbulence.

But in that last few decades much of this has changed and there has been a massive growth in the size of ground-based telescopes. This has been due to a technology called adaptive optics (see page 14). Thanks to this method we now have ground-based optical systems that can challenge some attributes of the space-based telescopes. Rather than being limited to space-based telescopes with mirrors 3-4m (10-13ft) in diameter, we have the wondrous ground-based leviathans set up today of

8-10m (26-33ft), with bigger telescopes planned for the near future.

What makes these amazing professional telescopes so great and how they are helping us learn more about the universe around us?

THE GEMINI SOUTH TELESCOPE

My favourite telescope on Earth today is the Gemini South telescope. As the name implies, there are two Gemini telescopes, collectively known as the Gemini Observatory. These twin telescopes are located on two of the best locations on the planet for observing the night sky: Hawaii, in the northern hemisphere and Chile, in the southern hemisphere. Each one has an impressive mirror 8.19m (26ft 10in) in diameter. Although these two telescopes don't have the largest mirrors in the world, they are definitely world-class, with cutting-edge technologies that allow scientists to see more and further. I had the thrill of working at Gemini South for around 18 months starting in 2003. It was a true joy for me, as I had made a telescope of around 150mm (6in) diameter as a child, so to be working on this amazing piece of engineering was a dream come true. My team and I had spent the previous three years building a new instrument for this telescope in a basement at University College London. The instrument itself did something quite special. It was a spectrograph: an instrument that takes light gathered by the huge telescope and splits it into its rainbow colours (spectrum). By analysing this light it is possible to work out whether a star is moving towards us or away from us and to establish where chemical reactions are taking place in the heart of a star. It was a magical time for me, so I always remember this telescope with absolute fondness.

SOLAR OBSERVATORIES

It may come as a surprise, but there are also dedicated telescopes used to observe the Sun. These solar observatories have been around for many years, with the first one founded in 1901 in Kodaikanal, India. As you can imagine with these daytime telescopes, they are designed with powerful solar filters and cooling systems. The study of space weather – the term used to describe the effect that the Sun has on conditions here on Earth – has become a large area of inquiry, due to the huge impact that solar events can have on some of our technology both on and in orbit about the Earth today. Solar activity of note include sunspots (dark spots that appear on the surface of the Sun), solar flares (violent bursts of radiation) and solar storms.

The solar storm that occurred in 1859, known as the Carrington event, for example, is one of the largest geomagnetic solar storms on record to date. Such storms occur when the sun sends out a giant cloud of magnetized matter (called a coronal mass ejection) that disturbs Earth's magnetic field. The Carrington event was so powerful that it took out much of the telegraph system in Europe and North America, and gave some telegraph operators electric shocks. It also enabled people to see the Northern Lights (aurora borealis) as far south as Hawaii and Cuba. If a similar storm were to hit with the same conditions today the effect on our satellites, communications and power transformers could be quite devastating.

Analysis of ice core samples have indicated that storms of this magnitude occur around once every 500 years, with smaller events of around a third of the magnitude occurring approximately three times per century.

Currently there are also around 10 active space solar observatories in orbit about the Earth. The Solar Dynamic Observatory (SDO) was launched in 2010 to help us understand how the Sun affects life on Earth. It observes the Sun in a range of wavelengths that are combined to work out what sort of solar activity is happening. The SDO website is very informative and displays a series of images of how the Sun looks at the moment.

SPACE TELESCOPES

As well as the solar space observatories, there are currently around 30 active space telescopes in operation, probably the most famous of which is the Hubble Space Telescope (HST). It is named after American astronomer Edwin Hubble (1889–1953), who was pivotal in giving us some of the first evidence that the universe is expanding. The HST, which is a reflector telescope about the size of an American large school bus, has been in operation for around a quarter of a century, and in that time it has transformed our understanding of the universe.

The HST did have some initial problems when scientists realised that one of its optical components had been manufactured to the wrong shape. After a very expensive fix, however the HST has gone on from strength to strength. Over the years it has beamed down breathtaking images from space that have given us an understanding of the age and size of the universe, the formation and life cycle of the stars and insight into some of the more exotic bodies out there, such as black holes (see page 38) and quasars (short for quasi-stellar radio sources; these are very distant and extremely bright objects that scientists believe are powered by black holes).

But to my mind the greatest thing that the HST has given us is a better view of our place in the universe. A series of pictures known as the Hubble Deep Fields has provided us with a clearer idea of how many galaxies there are in the universe. From data obtained by these images we now believe that there are around 100 billion galaxies in the universe. If each one of these galaxies has on average just 50 billion stars in them, and each of these stars has on average just two planets in orbit, you end up with... a feeling of how truly small we are in the scale of things.

It also gives us an indication of how likely it is that other forms of life may be out there.

10

THE FUTURE OF
STARGAZING

Amateur stargazing has gone from strength to strength in the last few years and I think that this trend will definitely continue into the future. As manufacturing processes continue to improve, although I don't believe that telescopes for amateurs will become cheaper, I think that we may get more for our money. The push for professional instrumentation with better CCDs (light-sensitive imaging devices; CCDs stands for 'couple-charged device') will eventually filter down to the commercial market, making larger and more efficient CCDs in telescopes available to the general public at relatively reasonable costs.

Many amateur users are even abandoning their eyepieces and instead using a CCD. These devices are the same as we have in our digital cameras. So rather than viewing an object directly you can use a CCD to give a live real-time image. The devices available at the moment are black and white, but they give us the ability to set up a telescope outside and then sit in the comfort of our houses and observe via the video acquired by the CCD. There are also some interesting processing techniques that can be used to, for instance, enhance dim images by increasing the gain (the amount of signal you get per photon of light).

CCDs are now used in professional astronomy rather than the old photographic plates as they produce a digital signal that can be easily stored and post-processed. Space telescopes use CCDs to transmit their observations and a digital signal is carried home on a radio wave.

As computer processing power and storage capacities get bigger and cheaper, I think that, as many amateurs are already doing, more and more of us will be enhancing the images we take to get the most out of the light we capture. So I predict that in the future adaptive optics systems may become available for amateurs, but unless commercial telescopes become a lot larger, around the 6-8m (19½-26ft) level, I think this may be a gimmick rather than a useful asset.

On the professional stargazing front, there are some fundamental questions that are still evading scientists and they are looking to build telescopes/instruments that will move them closer to the answers to deep questions, such as what is dark energy and dark matter; are there other dimensions, and are we alone in the universe? (Dark energy, which scientists think makes up about 68 per cent of the universe, is the mysterious energy that appears to be a key factor in the expansion of the universe; see page 38. Dark matter is an invisible matter that scientists think makes up about 28 per cent of the universe. Current measurements of the known, observed mass-energy content of the universe adds up to only 4 per cent – the rest of the universe, scientists surmise, is made up of dark energy and dark matter, but they just don't know what they are yet.) Scientists have got closer than ever with their understanding of the universe using current technology, but with the next generation of technology they may be able to find the answers that have eluded them. The quest continues in two core technologies: ground- and space-based telescopes and their instruments.

GROUND-BASED TELESCOPES

Since the introduction of adaptive optics to the telescope world, telescopes have been getting bigger and bigger and in the future they will continue to grow – we are truly living in exciting times. Radio astronomy (the study of radio signals from space), which can operate relatively unaffected by the atmosphere, is also thriving here on the ground, and there are some grand plans to transform this area of astronomy too. Radio astronomy allows us to peer into regions of space that are opaque to visible light. As mentioned in the previous chapter, looking at all parts of the electromagnetic spectrum gives us a much more detailed view of the processes occurring in the Universe.

The evolution in ground-based optical astronomy is best described through the ambitions of the scientists behind European Southern Observatory (ESO), a leading astronomy organisation in Europe. They were the master-builders behind the Very Large Telescope (VLT). This consists of four 8.2m ($27\frac{1}{2}$ft) telescopes located on the Cerro Paranal mountain in the Atacama Desert of northern Chile. This system and others like it are gathering data on the universe that competes well with some of our space-based telescopes, but the ESO team has not stopped there. They have now started construction on the European Extremely Large Telescope (E-ELT). This will be a multi-segmented mirror with a massive primary mirror of 39m (128ft) in diameter. This amazing beast will sit on the mountain next to its smaller VLT siblings in Chile and is due to start gathering light in 2024.

But the E-ELT is not the only new leviathan on the block. The Thirty Meter Telescope (TMT), a collaboration between universities in Canada and America, is due to come online in 2022 on the summit of Mauna Kea in Hawaii. The Grand Magellan Telescope (GMT) is under construction now, with a number of its seven 8.4m (27ft) mirror segments already constructed. This should see first light around 2021 at the Las Campanas Observatory, again in the Atacama Desert of Chile, so there should be some impressive science coming our way in the next few years.

But the ESO team has not just set its sights at extremely large; they want to go all out for 'overwhelmingly large' with the proposal for an Overwhelmingly Large Telescope (OWL). This colossus, if it is ever built, would have a segmented primary mirror of 100m (328ft) diameter – that is a light-gathering aperture roughly the size of a large football pitch. It remains a concept for now, but the things that astronomers would be able to see with such a wonderful telescope boggle the mind – undoubtedly, in the world of ground-based astronomy, people are thinking big!

Just as with amateur telescopes, professional ones are designed with different specialisms. The Larger Synoptic Survey Telescope (LSST) should be getting first light around

2019. As its name implies, this will be a survey telescope that will use the world's largest camera to photograph the whole of the night sky every few nights. To do this the telescope has been designed with a very large field of view. (The field of view is a huge 3.5° in diameter; to give you an idea of the scale, the Sun and the Moon are each a mere 0.5° in diameter when viewed from Earth). The primary mirror of the LSST will be 8.4m ($27\frac{1}{2}$ft), the first time that a mirror this size has been used on a survey telescope. Its main scientific goals make good use of its unique features by mapping our galaxy, the Milky Way, and looking for small, undetected objects in our solar system, especially asteroids that pass close to the Earth. LSST will also be looking for evidence of gravitational lensing (when light is bent by the gravitational field of an object in space) in the hope of detecting the signatures of dark energy and dark matter.

If we imagine that all of the objects that we see in the night sky sit on the imaginary celestial sphere, then it is possible to work out what sort of angle each object makes when we are observing it. Let's start locally, if you hold your thumb at arm's length and then your thumb subtends an angle of about 1° across. A fist subtends an angle of around 10°. In the same way when we look at objects in the sky they subtend an angle in the same way. The Sun and Moon happen to extend the same angle of 0.5° whereas the angle between the Asterism of the Plough and the Pole Star is around 30°.

RADIO TELESCOPES

There are already some huge radio telescopes in operation, such as the Arecibo Observatory in Puerto Rico, which has a diameter of 300m (984ft), and the Lovell Telescope at the Jodrell Bank Centre for Astrophysics in the UK with a diameter of 76.2m (250ft) – which is even more impressive when you consider it was completed in 1957. China's Five Hundred Meter Aperture Spherical Telescope (FAST) with a diameter of 500m (1,640ft) is under construction in Guizhou Province, southwest China, and should be active by late 2016.

As well as these large single-aperture radio telescopes, there is also a trend for large groupings of smaller radio telescopes. The biggest collection in existence at the moment is the Square Kilometre Array (SKA) project, which, when completed, will have thousands of radio dishes and over 1 million antenna located in Australia and South Africa. The combined collecting area of all of these dishes will be 1 square kilometre (1 million square metres/1,076,391 square feet). An array of this size will be the largest ever built and will be sensitive enough to detect an airport radar on a planet 10 light years away. One of the key science drivers behind the SKA is the search for and understanding of dark energy and dark matter (see page 38).

SPACE-BASED TELESCOPES

Although it is easier and much more cost-effective to build larger and larger ground-based telescopes, the space-based telescopes are not to be outdone.

Space-based telescopes have the advantage of being able to gather data in parts of the electromagnetic spectrum (see page 132-133) that ground based telescopes cannot access due to atmospheric absorption. Space-based telescope have been feeding our collective imaginations for a number of years now, with the Hubble Space Telescope leading the field (see page 138). Having been in space for over a quarter of a century, Hubble has transformed our understanding of the universe and given us images that brighten our hearts just by knowing such objects are out there. However, since the demise of the space-shuttle programme (NASA closed it in 2011), Hubble is on its own with no possibility of further extending its life now that shuttle astronauts will no longer be servicing it, so what will be launched in the next decade that will continue to feed our thirst for knowledge?

The most exciting development to date is the James Webb Space Telescope (JWST), which was originally called the Next Generation Space telescope (a name that has a nice *Star Trek* ring to it) and is due to launch in 2018. The telescope was renamed after NASA's second administrator, James E. Webb, which seems a strange thing to do until you realise that he was in this position from 1961 to 1968 and is the administrator who was responsible for getting the Apollo program literally off the ground (the first lunar landing took place the year after his retirement, in 1969).

Although JWST is thought to be the replacement for the Hubble Space Telescope, it will actually be working in a different wavelength. Hubble works from the visible to the ultraviolet and near-infrared ranges, while JWST has been designed to optimise observations in the infrared spectrum. This will allow scientists to see through the cloudy regions of space that only infrared light can penetrate, giving us new vistas into the universe. Due to the universe's expansion, many of the objects we observe have their light 'redshifted', i.e. the objects are so far away that the light they emit has been 'stretched' to the red end of the spectrum by the time it reaches our part of the universe. The light emitted from objects even further away that were formed early on in the universe fall into the infrared spectrum, so a telescope that can detect infrared will allow scientists to study some of the objects, such as galaxies, formed in the universe's infancy – a thrilling prospect.

The amazing telescopes mentioned above are just a small fraction of the exciting astronomical developments that are currently underway or that are being conceived. Professional astronomy is thriving, and this is great for all of us as scientists delve deeper and further out into the universe than ever before.

TOP 10 ASTRONOMERS OF THE ANCIENT WORLD

Astronomy is the science of the ancients and due to a plethora of amazing records we know of astronomers from over 4,000 years ago. Here is a list of notable astronomers who lived before 1000 AD in chronological order.

En-Hedu-Ana (c. 2285–2250 BC Akkad, now Iran)

En-Hedu-Ana was one of the first astronomers on record. She was the daughter of King Sargon I of Akkad, she was High Priestess of the Moon God. Amazingly we can understand the scope of her work through her poetry:

> "The true woman who possesses exceeding wisdom
> She consults a tablet of lapis lazuli
> She gives advice to all lands
> She measures the heavens
> She places the measuring-cords on the Earth."

Aristarchus of Samos (c. 310–230 BC Greece)

Aristarchus of Samos was a Greek astronomer who is the first person on record to talk about the heliocentric or sun-centred universe. In his book *The Sand Reckoner* he discusses that the universe was greater in size to what had been perceived and talks about the similarity between the stars in the night sky and the Sun. He also posits that the Sun (rather than the Earth) is the centre of the universe, that the Earth is in a circular orbit and that the stars are on a fixed sphere around the Sun. His ideas were rejected in favour of the geocentric universe proposed by Aristotle and Ptolemy.

Eratosthenes of Cyrene (c. 276–194 BC Libya/Greece)

Eratosthenes was a classic academic with interests in mathematics, astronomy, music and poetry. He is also thought to have invented the discipline of geography and made many calculations, including the distance between the Earth and the Sun. He invented the concept of the leap day, and calculated the circumference of the Earth and the Earth's tilt, all with amazing accuracy.

Hipparchus of Nicaea (Now Iznik, Turkey) (c. 190–120 BC Greece)

Hipparchus was one of the great early observationalists. He was able to work out the motions of the Moon and Sun using his own observations – and probably hundreds of years of data obtained by the Babylonians. He made some of the first star catalogues and also assigned stars magnitude according to their brightness.

Ptolemy (c. 90–168 AD Egypt/Greece)

Ptolemy wrote a number of influential treatises, including the *Almagest* which described the motions of the heavens using mathematics. In a later volume entitled *Planetary Hypothesise,* he proposed the model of the universe, later to be called the 'Ptolemaic System'. This was held to be the correct view of the universe across the world for around 1,200 years, until Copernicus reproposed the heliocentric model.

Aryabhata (c. 476–550 AD, India)

Aryabhata described a geocentric view of the solar system, ordering the objects out from the Earth and Moon: Mercury, Venus, the Sun, Mars Jupiter, Saturn and the constellations. He was able to come up with a scientific explanation for eclipses and calculate the length of a sidereal day and the sidereal year (a day and a year measured with reference to a fixed star) with outstanding accuracy. His year measurement was off by a mere three minutes and 20 seconds – less than 1000th of 1%.

Abū ʿAbdallāh Muḥammad ibn Mūsā al-Khwārizmī (c. 780–850 AD Persia)

Abū ʿAbdallāh Muḥammad ibn Mūsā al-Khwārizmī was born in the Islamic Golden Age. Another true all-rounder, he worked as a scholar in the House of Wisdom in Baghdad. He is probably most famous for his introduction of the decimal number system and algebra to Europe. As well as this he composed the Zij, astronomical books used to calculate the Sun's movement, the Moon and the five known planets of the time.

Al-battani (c. 850–929 AD Mesopotamia, now Turkey)

Son of a scientific instrument-maker Al-battani made improvements to some of the instruments available and was able to accurately calculate the length of a year. He also deduced that the Earth to Sun distance varied by observations of annular solar eclipses where the Sun's extent in the sky is much bigger than the Moon's. He was also able to calculate that the Earth tilted on its axis by around 23 degrees (remarkably close to the actual value).

Aḥmad ibn Muḥammad ibn Kathīr al-Farghānī (c. 880 AD Persia)

Another member of the house of wisdom, Abū al-'Abbās Aḥmad ibn Muhammed ibn Kathir al-Farghīnī or Alfraganus as he was known in the West – wrote astronomical books entitled *Sky Movements and the Science of Star Codes* and *The Theoretical Computations of Spheres*. These books were widely used by both Islamic and European scholars.

Abd al-Rahman al-Sufi (c. 903–986 AD Persia)

Abd al-Rahman al-Sufi was one of the most outstanding practical astronomers of the Middle Ages. He enhanced Ptolemy's original star charts by making corrections and including more details on their positions, magnitude and colour. His is also the first on record to have identified the Andromeda Galaxy and mention the Large Magellanic Cloud which he had either seen or knew of from reports. He also wrote about the astronomical instrument known as the Astrolabe which was generally used to measure the position of stars, but he included 1,000 additional uses for the instrument.

LOCATION OF NAKED EYE VISIBLE STARS WITH EXOPLANETS

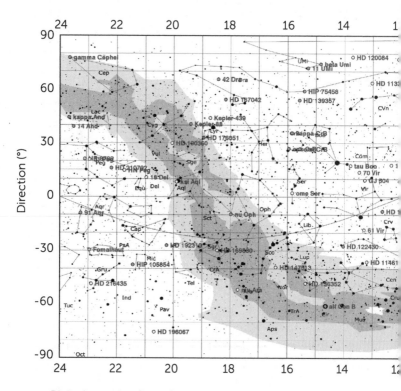

Right Ascension (hours)

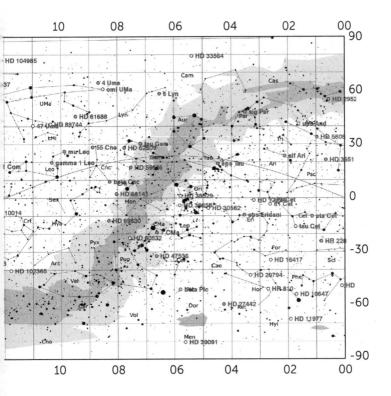

INDEX

ACKNOWLEDGEMENTS

Many thanks for the help and support of Claire, Romilly and all the guys from Quadrille Publishing and to my wonderful husband Martin, for saving me from my more obvious blunders.